Journal of Chromatography Library — Volume 11

LIQUID CHROMATOGRAPHY DETECTORS

JOURNAL OF CHROMATOGRAPHY LIBRARY

Volume 1 Chromatography of Antibiotics
by G. H. Wagman and M. J. Weinstein

Volume 2 Extraction Chromatography
edited by T. Braun and G. Ghersini

Volume 3 Liquid Column Chromatography. A Survey of Modern Techniques and Applications
edited by Z. Deyl, K. Macek and J. Janák

Volume 4 Detectors in Gas Chromatography
by J. Ševčík

Volume 5 Instrumental Liquid Chromatography. A Practical Manual on High-Performance Liquid Chromatographic Methods
by N. A. Parris

Volume 6 Isotachophoresis. Theory, Instrumentation and Applications
by F. M. Everaerts, J. L. Beckers and Th. P. E. M. Verheggen

Volume 7 Chemical Derivatization in Liquid Chromatography
by J. F. Lawrence and R. W. Frei

Volume 8 Chromatography of Steroids
by E. Heftmann

Volume 9 HPTLC —High Performance Thin-Layer Chromatography
edited by A. Zlatkis and R. E. Kaiser

Volume 10 Gas Chromatography of Polymers
by V. G. Berezkin, V. R. Alishoyev and I. B. Nemirovskaya

Volume 11 Liquid Chromatography Detectors
by R. P. W. Scott

Journal of Chromatography Library — Volume 11

LIQUID CHROMATOGRAPHY DETECTORS

R.P.W. Scott
Chemical Research Department, Hoffmann-La Roche Inc., Nutley, N.J.

ELSEVIER SCIENTIFIC PUBLISHING COMPANY
AMSTERDAM — OXFORD — NEW YORK   1977

ELSEVIER SCIENTIFIC PUBLISHING COMPANY
335 Jan van Galenstraat
P.O. Box 211, Amsterdam, The Netherlands

Distributors for the United States and Canada:

ELSEVIER NORTH-HOLLAND INC.
52, Vanderbilt Avenue
New York, N.Y. 10017

Library of Congress Cataloging in Publication Data

Scott, Raymond Peter William, 1924-
  Liquid chromatography detectors.

  (Journal of chromatography library ; v. 11)
  Includes bibliographical references and index.
  1. Liquid chromatography. I. Title.
II. Series.
QD79.C454S36   543'.08   77-23898
ISBN 0-444-41580-7

ISBN: 0-444-41616-1 (series)

ISBN: 0-444-41580-7 (vol. 11)

© Elsevier Scientific Publishing Company, 1977.
All rights reserved. No part of this publication may be reproduced, stored in a retrieval system or transmitted in any form or by any means, electronic, mechanical, photocopying, recording or otherwise, without the prior written permission of the publisher, **Elsevier Scientific Publishing Company, P.O. Box 330, Amsterdam, The Netherlands**

Printed in The Netherlands

Contents

Introduction IX

PART 1. GENERAL CHARACTERISTICS OF LIQUID CHROMATOGRAPHY DETECTORS

Chapter 1. History, function and classification of detectors. 1
    History and function 1
    Classification of detectors 3
    References 4
Chapter 2. Performance criteria of LC detectors 5
    The nature of the detector output 6
    Detector linearity 9
    The determination of the response index of a detector . . 11
        The dynamic range of a detector 13
        Detector response 14
    Detector noise 15
        Measurement of detector noise 17
    Detector sensitivity 18
    References 20
Chapter 3. Detector characteristics that affect column performance . 21
    Factors that directly affect band dispersion 21
        Theory 22
        Turbulence 26
    Factors that indirectly affect band dispersion 27
        The time constant of the recorder 33
        Adsorption effects in the detector 34
    Summary 35
    References 36
Chapter 4. Summary of detector criteria 37
Chapter 5. Ancillary equipment 41
    Predetector reactors 41
    Noise filters 45
    Integration 49
    References 53

PART 2. BULK PROPERTY DETECTORS

Chapter 1. General characteristics of bulk property detectors . . . 55
    The limitations of bulk property detectors 55
    General performance characteristics and areas of application 57
Chapter 2. The refractive index detector 59
    Theory 59
        The angle of deviation method 59

	The critical angle method	60
	The Fresnel method	60
	The Christiansen effect	62
	A commercial example of the refractive index detector	63
	Applications of refractometer detector	67
	References	67
Chapter 3.	The dielectric constant detector	69
	Theory	69
	References	78
Chapter 4.	The electrical conductivity detector	79
	A commercial example of an electrical conductivity detector	83
	References	90
Chapter 5.	Additional bulk property detecting systems	91
	The density detector	91
	The thermal conductivity detector	93
	The interferometer detector	95
	The density balance detector	98
	The vapor pressure detector	99
	References	104

PART 3. SOLUTE PROPERTY DETECTORS

Chapter 1.	Principles of detection	105
Chapter 2.	The ultraviolet absorption detector	109
	References	120
Chapter 3.	The fluorometric detector	121
	References	130
Chapter 4.	The polarographic detector	131
	References	138
Chapter 5.	The heat of adsorption detector	139
	References	148
Chapter 6.	The spray impact detector	149
	References	155
Chapter 7.	The radioactivity detector	157
	References	160
Chapter 8.	The electron capture detector	161
	References	166
Chapter 9.	Transport detectors	167
	References	179

PART 4. THE USE OF DETECTORS IN LIQUID CHROMATOGRAPHY

Chapter 1.	The selection of the appropriate detector	181
	The UV detector	181

Contents VII

	The refractive index detector	182
	The fluorescence detector	183
	The wire transport detector	184
	The electron capture detector	185
	The electrical conductivity detector	185
	References	186
Chapter 2.	Quantitative and qualitative analysis	187
	Column temperature	190
	Sample load	191
	Manual measurement of chromatographic data	192
	Computer data processing	193
	Qualitative analysis	195
	Quantitative analysis	197
	References	200
Chapter 3.	Practical hints on detector operation	201
	The recorder	201
	The amplifier	201
	Detector cells	202
	Bubbles — their removal and prevention	202
	Spurious peaks	203
	Base line instability	204
	Short term noise	204
	Long term noise	
	Drift	205
	The moving wire detector	205
	The conductivity detector	206
	The refractive index detector	207
Chapter 4.	Special detector techniques	209
	The differential detector	209
	The integral detector	215
	Vacancy chromatography	217
	References	221
Chapter 5.	Spectroscopic detectors	223
	The LC/UV combination	224
	The Variscan LC/UV spectrometer system	227
	The LC/MS combination	229
	LC/MS by direct sampling of column eluent	230
	The wire transport LC/MS system	233
	The Finnigan LC/MS transport system	240
	References	244
Index		247

INTRODUCTION

Liquid chromatography was the first chromatographic technique to be introduced and was employed for the separation of plant pigments as long ago as the latter part of the last century. Today it is a well understood, very effective and commonly used separation technique that is employed both by university and industry for analytical and preparative purposes. The development of the technique, however, was initially very slow and it is only in the last decade that real advances have been made. These advances have resulted in high efficiency columns, very sensitive detectors and instruments that can provide accurate quantitative and qualitative analyses. The rapid advancement of the technique over the past ten years has depended on the production of accurate high pressure pumps and well designed injection systems to provide minimum band dispersion on injection, but above all, it has depended on the development of high sensitivity in line detectors. The development of high sensitivity linear detectors permitted the accurate measurement of chromatographic parameters and thus paved the way for the modern, high efficiency, microparticulate packings.

This book discusses the important detector characteristics that affect both the quality of the chromatographic separation and the precision of the analytical results obtained. It describes how these characteristics can be measured and suggests optimal properties for specific chromatographic conditions. The most commonly used detector systems are discussed in detail, but the little known detection methods are also included, together with spectroscopic methods of detection. The purpose of the book is to give the readers a clear understanding of the principles of detection so that they may choose the most suitable detector for their purpose. It is hoped that the book will also stimulate research into developing improved or alternative detection systems and to encourage detector manufacturers to improve their designs, to permit the capabilities of present day high efficiency columns to be fully realized.

The author would like to thank his colleagues, Dr. C. G. Scott, Mr. P. Kuocra and Mr. C. Reese for their many helpful suggestions in the preparation of this book and Miss H. Rennie for typing the manuscript.

R. P. W. Scott
20 June 1977

PART 1

GENERAL CHARACTERISTICS OF LIQUID CHROMATOGRAPHY DETECTORS

CHAPTER 1

History, Function and Classification of Detectors

History and Function

A liquid chromatographic detector is a device that locates, in the dimension of space or time, the positions of the components of a mixture that has been subjected to a chromatographic process and thus permits the senses to appreciate the nature of the separation that has been obtained. This definition, by necessity, has to be broad as it needs to encompass all detecting systems ranging from the elaborate electronic devices presently available, to the human eye or the sense of smell. Tswett in his pioneering chromatographic separation of plant pigments used the human eye to determine the nature of the separation he obtained and even today, as the most common separation technique employed is thin layer chromatography, the human eye is still the most frequently used detector. The human eye, as a liquid chromatography detector, however, has severe limitations. The majority of substances that are chromatographed are colorless and thus have to be chemically changed to render them visible; further the retinic response of the eye is not linear which, when coupled with the variation of the iris to light intensity, makes the eye a poor detecting system for quantitative estimation. The human eye, in fact can only be used for quantitative work as a null sensing device where closely similar light intensities are being matched as in the use of a comparator.

    The lack of a satisfactory detector was probably the greatest single impediment to the development of liquid chromatography over the years past. The rapid development of gas chromatography arose solely from the availability of sensitive detecting systems and it is interesting to note, that liquid chromatography had to await the introduction of sensitive liquid chromatography detectors, before the rapid advances in the technique that have occurred over the past few years, could take place. High sensitivity detectors have provided accurate concentration profiles of eluted solutes and allowed the magnitude of the solute dispersion (band spreading) that occurs in the column to be determined. The data so produced has permitted the development and confirmation of theories that describe the various factors that contribute to band dispersion which in turn have improved column technology. Thus, the impressive separations that are achieved today in both gas and liquid chromatography can be directly attributed to the introduction of sensitive on-line detecting systems; that is detectors that can be directly connected to the column outlet.

The first attempt to develop an alternative to visual detection was to collect the column eluent as a large number of fractions and to subsequently analyze each fraction by appropriate techniques such as colorimetry or titration. The concentration of solute in each fraction was then plotted against fraction number and a chromatographic histogram obtained. This procedure was extremely laborious and time consuming and was only effective for well resolved mixtures. Further a chromatographic histogram did little to aid in the improvement of column technology. The concept of on-line detection, where an appropriate sensing device is connected directly to the column outlet, was established in the late thirties and early forties. Such a device provides a suitable signal, that is directly related to the concentration of solute in the eluent. Examples of two of the early on-line detectors are given by the conductivity detector described by Martin et al (1) and the refractive index detector described by Tiselius (2). This book is concerned solely with on-line detection and as the original definition of liquid chromatography detectors embraced all kinds of detecting methods it would be appropriate to define specifically, on-line detectors. An on-line liquid chromatography detector is a device that provides a continuous output, usually electrical in nature, that is some function of the mass of solute or concentration of solute in the mobile phase as it leaves the column. The output from such a system can be passed directly, or by way of some suitable electronic device, to a strip chart recorder which will then provide a continuous record of the mass of solute eluted, or concentration of solute in the eluent with respect to time. Such systems can provide accurate concentration profiles of eluted solute bands and thus a quantitative analysis of the original mixture can be obtained.

The major progress in the development of liquid chromatography has occurred over the past two decades and has been due in part to the impetus provided by the rapid advances that have taken place in the technique of gas chromatography (GC). Gas chromatography detection methods have been directly applied to the development of liquid chromatography detectors and furthermore, many of the scientists involved in the development of GC detectors, on exhausting the challenges of gas chromatography, turned their attention to the field of liquid chromatography (LC). Between 1956 and 1960 at least six GC detectors were invented and developed to their full potential; the progress in the development of LC detectors however has been in contrast slow and arduous.

The problem of detection in liquid chromatography is far greater than in gas chromatography because low concentrations of solute in a liquid do not modify the overall physical characteristics of a liquid (e.g. density, dielectric constant, etc.) to the same extent, that low concentrations of vapor in a gas modify the characteristics of the gas. It follows that LC detectors which function on the principle of measuring some overall physical property of the eluent will be very very sensitive to changes in ambient conditions and fluctuations in column flow rate and consequently exhibit very limited sensitivity. Nevertheless steady progress in detector development has been made and there are now a number of effective LC detectors available. However there is, at present, no liquid chromatography detector that possesses all the necessary attributes that are required for a completely versatile liquid chromatograph. There is no universal detector in liquid chromatography that compares with the flame ionization detector in gas chromatography. The most generally sensitive detector is the U.V. detector and the most versatile probably the wire transport detector.

Classification of Detectors

Detectors can be broadly classified into two types. <u>Bulk Property</u> detectors which function by measuring some bulk physical property of the column eluent (e.g. dielectric constant or refractive index) and <u>Solute Property</u> detectors which function by measuring a physical and/or chemical property that is characteristic of the solute only (e.g. UV adsorption). This classification is not completely precise, for example the UV detector which is classed as a solute property detector when used with an ethyl acetate-heptane mobile phase will give a constant background signal due to the UV adsorption of the ethyl acetate. It follows that the UV detector, although a solute property detector, under some condition of use, behaves as a hybrid between a bulk property detector and a solute property detector.

Irrespective of its class, ideally a liquid chromatography detector should have the same characteristics as a gas chromatography detector with a sensitivity of about 10^{-12}-10^{-11} g/ml and a linear dynamic range of six orders. The liquid chromatography detector should also be completely versatile and detect all the solutes while at the same time being independent of the characteristics of the mobile phase. Under such circumstances changes can be made in the composition of the mobile phase during development, for example, by gradient elution. Unfortunately no liquid chromatography detector so far devised nearly approaches the above specifications. However, if taken together as a group, the present liquid chromatography detectors do embrace most of the important characteristics of the ideal detector. For this reason present work in the field of liquid chromatography often needs a number of detectors of different types to be available, so that the appropriate one can be chosen for any particular application.

REFERENCES

1. A. J. P. Martin and S. S. Randall, Biochem. J., 49 (1951) 293.
2. A. Tiselius and D. Claesson, Arkiv. Kemi Mineral. Geol., 15B (No. 18) (1942).

CHAPTER 2

Performance Criteria of LC Detectors

In order to evaluate a detector for use in liquid chromatography, performance criteria or specifications must be available to assess its pertinence to particular chromatographic separations and also to permit a rational comparison with other detectors. It follows that performance data needs to be provided for each detector, that is presented in a standard form and given in standard units which will be consistent between detectors that function on widely different principles. The principal characteristics of a detector that will fulfill these requirements are Response, Linearity, Noise and Sensitivity. There is much confusion between liquid chromatographers, not only with respect to the units in which the above specifications should be given, but even in the exact definition of these criteria. This confusion has resulted partly from the use of criteria developed for other instrumental devices which are not applicable to LC detectors and partly due to some manufacturers selecting ambiguous criteria in order to present their products in the best possible light. It follows that the various criteria and specifications necessary to describe the properties of a given detector must be described in detail and clearly defined.

A detector can be either a mass sensitive or concentration sensitive device, that is to say the detector output can be some function of mass of solute passing through it per unit time or mass of solute per unit volume of mobile phase. The flame ionization detector used in GC is a mass sensitive detector in that, within limits, the response is constant for a given mass flow and independent of the volume flow of hydrogen or carrier gas that passes through it. LC detectors are mostly concentration sensitive devices, with the possible exception of the mass detector and thus the parameter measured by the detector will be assumed to be mass/unit vol or g/ml. The wire transport detector contains a flame ionization detector but as will be shown later, this detector is still a concentration sensitive device. Although in the discussion of different detector criteria, it will be assumed that the detector response is some function of solute concentration (g/ml), alternative equations are given in terms of g/sec should mass sensitive detectors be developed in the future.

The Nature of the Detector Output

There are basically three types of detector output - linear, integral and differential. A linear detector, as its name implies, is one that provides a signal that tends to be directly proportional to the concentration of solute in the mobile phase passing through it.

$$\text{thus } y = Ac$$

where y is the output of the detector in appropriate units

c is the concentration of solute in the mobile phase in g/ml

and A is a constant.

All detectors are designed to give as near a linear response as possible to minimize calibration and calculation procedures when used for quantitative analysis. In many instances the output from the sensing device of the detector may not be linear, in fact it is often logarithmic or exponential and thus the associated electronics have to be designed to convert the output to a linear response. For example a sensing device with an exponential output when used with a logarithmic amplifier would provide a linear response.

Theoretically, there is no reason why a detector should not have a logarithmic, exponential or any other functional output provided the function can be explicitly defined. However, this would render chromatograms from the detector difficult to interpret, and quantitative analysis more involved and time consuming. For this reason detectors that do not have a linear output are not readily acceptable to chromatographers generally. The profile of an eluted peak closely resembles the Gaussian or error function curve where the independent variable is volume of mobile phase and the dependent variable is solute concentration. If the flow rate through the column is kept constant, then the independent variable can be replaced by time. Thus, a linear detector will provide an accurate representation of the Gaussian profile of the peak as shown in figure 1.

If a detector with a linear output is connected to an integrating analog amplifier, an integral output can be simulated, an example of which is also given in figure 1.

Nature of the Detector Output

For a concentration sensitive detector to directly provide an integral response.

$$y = \int_0^v c\,dv = cv = m$$

where y is the detector output
c is the concentration of solute/unit volume
v is the total volume of mobile phase passed through the detector
m is the total mass of solute

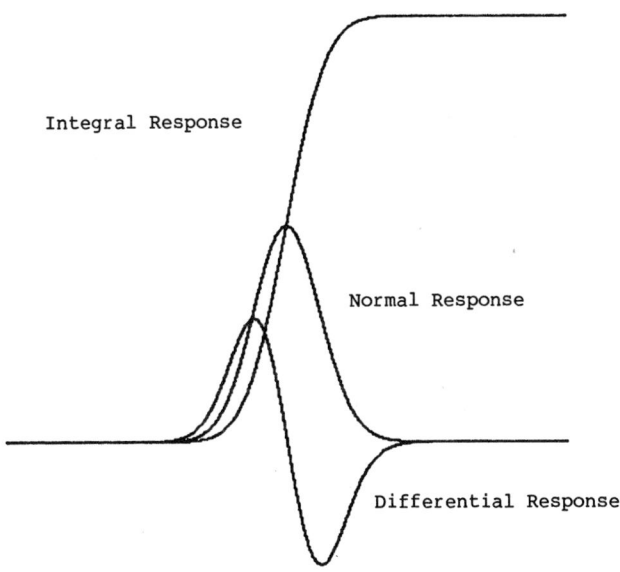

Figure 1 Types of Detector Response to a Gaussian Elution Curve

For a mass sensitive detector then

$$y = \int_0^t c'\,dt = c't = m$$

where c' is the mass of solute/unit time passing through the detector and t is the time

It is seen that in both cases the integral curve is directly related to mass of solute, and that the step height of an integral curve is proportional to the total mass of solute eluted.

To the best of the author's knowledge only one detector in normal operation provides an integral output and that is the mass detector. However, all detectors provided with flow-through cells or the equivalent can, if desired, be used to provide an integral output as will be discussed later. The integral chromatogram has advantages for quantitative analysis providing the peaks are well resolved; the integral chromatogram, however, is not popular among most chromatographers as, where separation is not complete, and for multicomponent mixtures, the individual solute bands are less easily located and identified than in the normal chromatogram.

In a like manner, a differential output can be simulated if the output from a linear detector is connected to a differential analog amplifier and an example of such a curve is also included in figure 1.

For a concentration sensitive detector which provides directly a differential output

$$y = \frac{dc}{dv}$$

and for a mass sensitive detector

$$y = \frac{dc'}{dt}$$

The value of the differential curve is very limited as the chromatogram having unresolved peaks becomes complex and difficult to interpret. It is seen from figure 1 that at the normal peak maximum the differential curve goes from a positive value, through zero to a negative value thus the electrical analog of this signal can be made to activate a suitable timing device for the automatic recording of retention times. In normal operation the only detector that can give a differential response is the heat of adsorption detector and then only under very specific conditions. Under most working conditions the heat of adsorption detector response only tends towards the differential form, in that the output for a normal Gaussian peak will be sigmoid in shape, the front being a positive peak which is followed by a negative tailing peak.

Detector Linearity

As stated, a linear detector is one with an output that is described by the following equation:

$$y = Ac$$

Where y, A and c have the meanings already defined.

Due to the imperfections of electrical and mechanical devices no detector is truly linear and designers of detectors can only try to approach this ideal function. Most manufacturers claim that their detectors are linear, but the extent to which their instruments approach true linearity varies significantly between one and another.

It is therefore important for a chromatographer to have some measure of the linearity of the detector in use and this requires some means of providing a numerical value that indicates the extent to which the function deviates from true linearity.

Fowlis and Scott (1) suggested a way in which a measure of the detector linearity could be determined. They put forward the following power function that could accurately describe the output of all detectors that were approximately linear.

$$y = Ac^r \qquad (1)$$

Where r is the Response Index of the detector and the other symbols have the meaning previously ascribed to them.

It follows that for a truly linear detector $r = 1$ and the proximity of r to unity would indicate the extent to which the detector's performance deviates from true linearity. Figure 2 shows curves relating detector output to solute concentration for different values of r. It is seen that all the curves approximate closely to a straight line, but the errors involved in assuming that they are linear are shown in Table I.

It is seen from Table I that errors in the lower level component can be as much as 12.5% for $r = 0.94$ and 9.5% for $r = 1.05$. For most practical purposes r should lie between 0.98 and 1.02 if reasonable linearity is to be assumed. However, it should be pointed out that if r is known, then a correction can be applied, and thus take into account any non-linearity that may exist. It should also be emphasized that for the most accurate work the response index of the detector should be determined and applied where appropriate.

Table I

THE ANALYSIS OF A TWO COMPONENT MIXTURE USING DETECTORS
HAVING DIFFERENT RESPONSE INDICES

Solute	r = 0.94	r = 0.97	r = 1.0	r = 1.03	r = 1.05
1	11.25%	10.60%	10%	9.42%	9.05%
2	88.75%	89.40%	90%	90.58%	90.95%

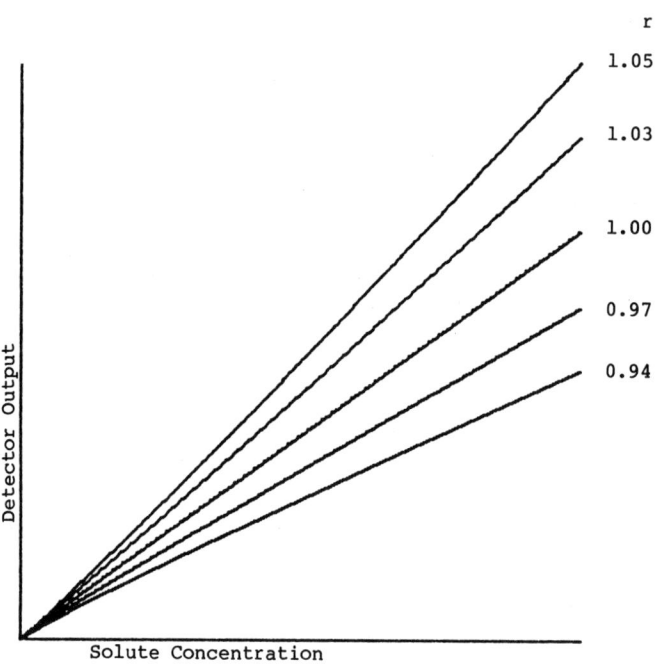

Figure 2 Graphs of Detector Output against Solute Concentration for Detectors having Different Response Indices

The Determination of the Response Index of a Detector

There are two methods that can be used to measure the response index of a detector, the Incremental Method and the Logarithmic Dilution Method. The former requires no special apparatus other than the chromatograph itself, while the latter requires special apparatus which fortunately is very simple to fabricate. The incremental method employs the detector, together with its ancillary electronic apparatus and recorder, connected to a suitable column. The column can be packed with glass beads so that sufficient band spread is produced to allow a measurable peak to be eluted. Depending on the absolute sensitivity of the detector, samples of the chosen solute of appropriate size are dissolved in the mobile phase, and injected onto the column and the peak recorded. Duplicate injections are recommended for each solute concentration. The sample size for successive calibrations should be increased by a factor of three until the sensitivity range of the detector is covered. The width of each peak is measured at 0.607 of the peak height (3), and from the chart speed and the mobile phase flow rate, the peak width in ml of mobile phase is calculated. It should be noted that the peak width at 0.607 of the peak height is equal to twice the standard deviation of the peak, which is assumed to be Gaussian. The distance measured is thus equivalent to half the peak width at the base. It follows that from the knowledge of the mass of solute injected and the width of the base of the peak in ml of mobile phase, the average concentration of the solute in the total peak can be calculated. Now the concentration at the peak maximum can be taken as twice the average concentration, and so the maximum concentration of the peak can be calculated from the following equation:

$$c = \frac{ms}{wQ}$$

Where c is the concentration of solute in the mobile phase at the peak maximum in g/ml
m is the mass of solute injected in g
w is the peak width at 0.607 of the peak height in cm
s is the chart speed of the recorder in cm/min
and Q is the flow rate in ml/min

The logarithm of the peak height (y) is then plotted against the logarithm of the solute concentration at the peak maximum (c) as given by equation 2.

From equation (1):

$$\text{Log } y = \text{Log } A + r \text{ Log } c \qquad (2)$$

Thus the slope of the Log/Log curve will give the response index r. If the detector is truly linear, $r = 1$, and the slope of the curve will be $\pi/4$.

The logarithmic dilution method for detector calibration was introduced by Lovelock (2) for gas chromatographic detectors and was further modified by Fowlis and Scott (1) for liquid chromatographic detectors. The system provides a continuous flow of solvent containing solute, the concentration of which decreases logarithmically with time.

A known mass of solute is introduced into a well-stirred reservoir through which a flow of pure solvent continually passes. The solution is thus continually diluted and the concentration of the solute in the exit flow from the reservoir is monitored by the detector under examination. A diagram of the dilution system is shown in figure 3. Let the vessel have a volume V and let the concentration of solute in the vessel be c_t after time t. Let a volume dv of pure solvent enter the vessel displacing a volume dv from the vessel. The mass of solute removed will be $dm = c_t \, dv$. Now the change in mass dm in the dilution vessel will result in a change of solute concentration of dc_t.

$$\text{Thus } V dc_t + c_t dv = 0$$

$$\text{and } \quad \frac{dc_t}{c_t} = \frac{-dv}{V}$$

$$\text{Integrating Log } c_t = \frac{-v}{V} + k$$

where k is the constant of integration

Now $v = Qt$ where Q is the flow rate through the vessel

$$\text{Thus Log } c_t = \frac{-Qt}{V} + k$$

Now when $t = 0$, $c = c_o$ where c_o is the initial concentration of solute in the dilution vessel.

$$\text{Thus } k = \text{Log } c_o$$

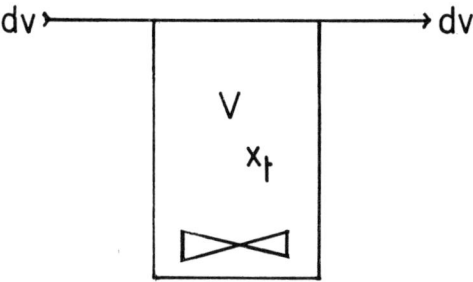

Figure 3   Diagram of Logarithmic Dilution Vessel

$$\text{Therefore Log } c_t = \frac{-Qt}{V} + \text{Log } c_o$$

$$\text{or } c_t = c_o e^{\frac{-Qt}{V}}$$

Thus if the logarithm of the detector output is plotted against time then for a truly linear detector a straight line will be produced having a slope of $\frac{-Q}{V}$. If the detector has a response index of r and the slope of the line is ϕ

$$\text{Then } \phi = \frac{-Qr}{V} \quad \text{or} \quad r = \phi \frac{V}{Q}$$

Thus the response index of the detector can be determined, but the accuracy of this determination will depend upon the constant nature of Q, the flow rate, and so a good quality constant flow pump should be employed. Manufacturers generally do not give the response index of their detectors, and, therefore, for highly accurate work its value needs to be determined.

The Dynamic Range of a Detector

There are two quite different response ranges of a detector which the operator should be aware. The Dynamic Range of the detector and the Linear Dynamic Range of the detector, and in most instances the two ranges are not synonymous. The dynamic range of the detector is that solute concentration range over which the detector will provide a concentration dependent output. The minimum of this range will be at the solute concentration level where the detector output is twice the value of the noise (which will be discussed later), and the maximum where the output of the detector fails to respond to an increase in solute concentration and becomes saturated. The dynamic range is usually quoted in orders of magnitude of solute concentration and is given the symbol D_R; eg, $D_R = 2 \times 10^{-6}$ to 4×10^{-3} g/ml.

The linear dynamic range of a detector is that range of solute concentration over which the output of the detector is linear within a given response index range. The linear dynamic range of a detector is not the same as its dynamic range as already stated, as the linearity of most detectors deteriorates at high concentration of solute and in some instances also at very low concentrations of solute. The linear dynamic range of a detector is also quoted in order of magnitude of concentration and is given the symbol D_L; eg, $D_L = 5 \times 10^{-8} - 2 \times 10^{-5}$ g/ml ($0.98 < r < 1.02$). At present manufacturers do not usually differentiate between D_R and D_L and do not quote a range for the response index r, however it is hoped that as the technique for LC develops such data will be made available. Some manufacturers do mark the least sensitive setting of the detector as non-linear (N/L), which is a step towards a more rational approach to specifying dynamic range.

Detector Response

The detector response is defined in two ways depending on whether the detector is mass sensitive or concentration sensitive. For a mass sensitive detector the response is given in mv/mass/unit time and has the symbol R_m and for a concentration sensitive detector the response is given in mv/mass/unit vol having the symbol R_c. Assuming the mass/per unit time or mass/per unit volume of solute eluted at the peak maximum is twice the average mass/unit time or mass/unit volume eluted throughout the entire peak then

$$R_m = \frac{hw}{sm} \text{ and } R_c = \frac{hwQ}{sm}$$

Where h is the peak height in mv

w is the peak width at 0.607 of the peak height in cm

m is the mass of solute injected onto the column

s is the chart speed in cm/min

Q is the flow rate of mobile phase in ml/min

The response for a given detector will be different for different solutes; in the case of the UV detector, the response will be a function of the extinction coefficient of the solute and for a refractive index detector, the refractive index of the solute. For this reason the response of two detectors of the same type and geometry can only be compared if the same solute and mobile phase are employed. When comparing the response of detectors of the same type but different geometry then other factors have to be considered as well. For example, in comparing the response of the two UV detectors from different manufacturers the path lengths of the respective detecting cells must be taken into account as indicated by Beers law which will be discussed later.

Detector Noise

Detector noise is the term given to any perturbation on the detector output that is not related to an eluted solute. Detector noise is an extremely important characteristic of a detector as it determines the ultimate sensitivity of the device. Detector noise has been arbitrarily divided into three types which are depicted in figure 4. The first type is short term noise sometimes called "grass", and consists of base line perturbations that have a frequency that is significantly higher than that of the eluted peak. Short term noise usually arises from the detector or recorder electronics. This type of noise is not usually a serious problem as it often can be entirely eliminated by an appropriate noise filter, and further does not seriously obscure the presence of a solute peak as shown in the upper curve of figure 4. The second form of noise is called long term noise and consists of base line perturbations that have similar frequency to the eluted peaks. This type of noise is shown in the second curve in figure 4. This is the most serious type of detector noise as it cannot be differentiated from an eluted peak of the same amplitude. The eluted peak that can be easily identified in the first curve cannot be identified from the long term noise in the second curve. Furthermore, any filter that would eliminate long term noise would virtually remove the solute peak as well.

Long term noise arises largely from the detector itself caused by instabilities of its components or its susceptability to small changes in ambient conditions.

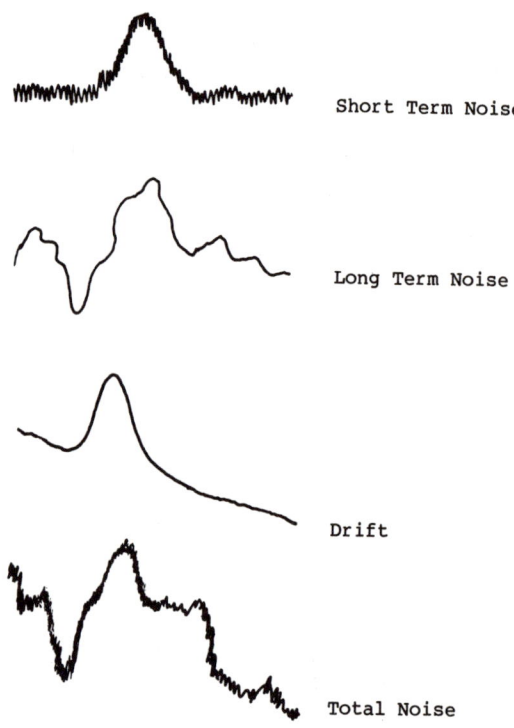

Figure 4   Different Types of Detector Noise

In some instances long term noise may depend on variables to which the particular detector is sensitive. For example the refractometer detector, sensitive to temperature changes is also sensitive to pressure changes as the refractive index of liquids vary with pressure. It follows that the fluctuation of pressure across the refractometer cell resulting from changes in column flow rate will affect the magnitude of any long term noise. Such noise can only be reduced by improved design of the detector.

Measurement of Noise

Perturbation of the detector output having a frequency significantly greater than the frequency of the eluted peak is called drift and an example is shown in the third curve in figure 4. It is seen that again the drift does not obscure the eluted peak but detectors operating with significant drift will require frequent adjustment of the base line. Drift can result from slowly changing output from the power supply to the detector, changes in ambient temperature, but more often is due to changes in the composition of the column eluent resulting from incomplete equilibrium being attained on changing the mobile phase. When detectors are operated at or near maximum sensitivity, all three types of noise are usually present and the detector output resembles the fourth curve in figure 4.

Measurement of Detector Noise

The quantitative assessment of detector noise is taken as the maximum amplitude of the combined short and long term noise measured over a period of about ten minutes. The detector must be connected to the column and the mobile phase passed through it over the period of measurement. For a 4-5 mm I.D. column a flow rate of 1 ml/min is appropriate and the maximum amplitude of the combined short and long term noise measured in mv and corrected to an attenuation of unity as shown in figure 5. The detector noise will be given the symbol N_D.

$$N_D = v_1 A = v_1/S$$

where v_1 is in millivolts as measured on the recorder scale
A is the attenuation factor
and S is the amplification factor

It should be noted that attenuation is the reciprocal of amplification and individual manufacturers may use either method to designate the sensitivity setting

| Minimum Sensitivity | Amplification Attenuation | 1 32 | 2 16 | 3 8 | 8 4 | 16 2 | 32 1 | Maximum Sensitivity |

Where detectors are sensitive to pressure changes and variation in column flow rate the noise level under static conditions is sometimes quoted. This is often the method for specifying noise level for refractometer detectors which are very sensitive to pressure changes within the cell

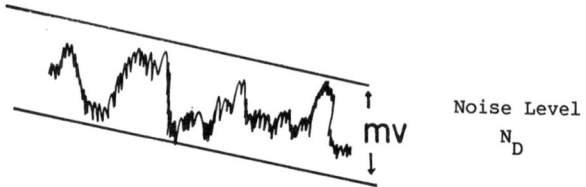

Figure 5   Measurement of Detector Noise Level

resulting from variations in flow rate. Although useful, such specifications of noise level are not really valid to the chromatographer as he can never utilize the detector without a flow of mobile phase passing through it. It could be argued that the manufacturer of detectors cannot be responsible for the pump providing constant pressure or a constant flow rate. However, all solvent delivery systems provide mobile phase flow rates that have some variation and it is really the responsibility of the detector manufacturer to design his detector such that the effects of pressure and flow variations are minimized.

It should be noted that at the high sensitivity ranges of some detectors, filter circuits are automatically introduced to reduce noise. Under these circumstances the noise level should be determined at the lowest attenuation (or highest amplification) that does not include noise filtering devices and then corrected to an attenuation of unity.

Detector Sensitivity

Detector sensitivity is the minimum mass/unit time or minimum concentration of solute in mass/unit volume passing through the detector that can be discerned from the noise. The size of the signal relative to the noise that can allow the solute signal to be discerned from the noise has to be arbitrarily defined and it is generally accepted that a signal to noise ratio of two will permit unambiguous identification of a signal. It follows that for a detector responding to the mass passing through it per unit time the detector sensitivity S_m is given by

$$S_m = \frac{2N_D}{R_m} \text{ g/sec}$$

and for a concentration sensitive detector the sensitivity S_c is given by

$$S_c = \frac{2N_D}{R_c} \text{ g/ml}$$

R_m, R_c, N_D being determined in the manner previously described. It should be emphasized that the sensitivity of a detector is <u>not</u> the minimum <u>mass</u> of solute that can be detected. During the development of a chromatogram the peaks become broader as the retention volume increases. Thus a given mass of solute may be detected at low retention where peaks are sharp and high, but if the chromatographic conditions are changed, so that the same mass of solute is eluted later in the chromatogram, the peak will be broad and low and may not be discernible from the noise. Thus any sensitivity quoted as a minimum mass detectable must be carefully examined and if the data is available the sensitivity must be calculated in the manner given above. Some manufacturers have taken the minimum detectable concentration and multiplied its value by the cell volume of the detector and given the product as the minimum detectable mass of solute. This method of specifying detector sensitivity is particularly misleading. For example a concentration sensitive detector having a true sensitivity of 10^{-6} g/ml and a cell volume of 10 µl will be attributed a sensitivity of 10^{-8} g. This of course is grossly incorrect as the detector volume must always be at the most 1/20 of the base width of the peak so under ideal conditions the minimum mass of solute that can be detected will be about 10^{-5} g or 10 µg. Thus, even when arranging the chromatographic conditions to provide the maximum apparent mass sensitivity, the values quoted on the basis of the product of the true sensitivity and the cell volume will be over 3 orders in excess of the real value. Specifications for detectors given by manufacturers are becoming more rational, but misleading values for detector sensitivity are still given and chromatographers need to carefully assess their significance if not given in the form discussed here.

Sensitivity is one of the most important parameters of a liquid chromatographic detector. Liquid chromatography adsorbents and particularly bonded phases have very low loading capacities, that is, only small charges can to be placed on the column, and therefore to realize the full resolving power of the column, detectors of high sensitivity must be employed. In fact the most sensitive detector, the UV detector, has to be used with columns packed with bonded phases or efficient separations of trace components will not be possible. Another important, but less well known effect of detector sensitivity on the performance of a liquid chromatograph column is its effect on peak capacity (4). The later the peak is eluted in the chromatogram the more dispersed it becomes, and the lower the concentration of the solute at the peak maximum. Peaks will be eluted and detected until the concentration of the peak maximum is reduced to the limiting sensitivity of the detector. Solutes may well be eluted subsequent to this point, but will never be detected. It follows that the useful length of the chromatogram, that is the number of solutes that can be eluted, is ultimately limited by the sensitivity of the detector employed.

REFERENCES

1. I. A. Fowlis and R. P. W. Scott, J. Chromatogr., 11 (1963) 1.
2. J. E. Lovelock, Gas Chromatography, (Ed R. P. Scott), Butterworth, London, 1960, p. 26.
3. Bonded Stationary Phases in Chromatography, (Ed E. Grushka), Ann Arbor Science, Ann Arbor, Mich., 1974.
4. R. P. W. Scott, J. Chromatogr. Sci., 9 (1971) 449.

CHAPTER 3

Detector Characteristics that Affect Column Performance

Factors That Directly Affect Band Dispersion

Separations are achieved in liquid chromatography by employing a mobile and stationary phase system that will move the individual solute bands apart during development and by designing the column to keep the individual solute bands narrow. Obviously, the more narrow the bands and the further they are moved apart, the better the separation. The detector and its connecting tubes cannot affect the degree to which the solute bands are moved apart, as this depends solely on the characteristic of the two phases, but they can affect the width of the solute bands. Band spreading in the connecting tubes or cell volume itself results from poor radial transfer of the solute in the liquid and the parabolic velocity profile of the mobile phase that exists in the connecting tubes and the detector cell itself. As the technique develops, columns will provide higher and higher efficiencies, which means the bands will become more narrow and provide improved resolution. At present the base width of the peak eluted from a high efficiency column in terms of volume of mobile phase may only be about 100 microliters, and it follows, therefore, that a detector having a cell volume of 8 microliters will significantly contribute to the band width and thus the real efficiency and resolving power of the column not realized. Peaks already separated in the column will be merged together due to the solute bands being broadened in the detector. In a similar way the time constant of the amplifier and the recorder can contribute to the apparent band width as recorded on the chart. Most amplifiers and recorders have filter circuits associated with their outputs to eliminate high frequency electronic noise, but if the filter frequency limit approaches that of the eluted peak then the peak will be reduced in height and broadened. As the design of the detector and its electronics can significantly affect the performance of the column with which it is associated those characteristics

Theory

It is not the intent of this book to give a treatment of the theory of chromatography and for those wishing to study the subject more fully are recommended to read the appropriate chapters in <u>Contemporary Liquid Chromatography</u> by the same author and published by J. Wiley. However, the basic principles of the theoretical approach that must be used to investigate dispersion in detector cells and detector connecting tubes will be outlined and the pertinent equations given.

The dispersion processes that occur in the column connecting tubes and detector cells are all random in nature and therefore broaden any Gaussian distribution of solute concentration but maintains its Gaussian form. Thus the Gaussian curve relating solute concentration to volume flow of mobile phase or time as measured by the detector is made up of the elution curve resulting from the dispersion effects in the column, together with the curve resulting from the dispersion effects occurring in the connecting detector tubes and the detector cell itself. It can be shown that the variances of each dispersion effect can be summed to give the variance of the final curve T^2.

Thus

$$T_C^2 + T_T^2 + T_D^2 = T^2$$

where T_C is the standard deviation of the solute band eluted from the column

T_T is the standard deviation of the curve resulting from dispersion in the connecting tubes

and T_D is the standard deviation of the curve resulting from dispersion in the detecting cell

It is generally accepted that the band width of a peak can be increased by 5% (1) without seriously impairing the efficiency and resolution of the column.

$$\text{Thus } T = 1.05 \, T_C$$

$$\text{Hence } T_C^2 + T_T^2 + T_D^2 = (1.05 T_C)^2$$

$$\text{or } T_T^2 + T_D^2 = 0.103\, T_C^2$$

Thus the sum of the variances resulting from dispersion in the connecting tubes and the detector cell should not exceed 10% of the variance of the band leaving the column. Scott and Kucera (2) examined the effect of band dispersion in capillary tubes and derived the following expression:

$$r = \left(\frac{2.4\, D_M V_R^2}{\pi n Q l}\right)^{1/4}$$

Where r is the radius of the tube in cm

D_M is the diffusivity of the solute in the mobile phase in cm^2/sec

V_R is the retention volume of the solute in ml

n is the efficiency of the attached column in theoretical plates

Q is the volume flow rate through the column in ml/sec

and l is the length of the capillary tube in cm

This equation can be used to calculate the dimensions of a detector cell or the connecting tubes for a column of given dead volume and efficiency. However, the detector comprises of a cell and connecting tubes, and, therefore, both must be taken into account. As the detector cell is the most important part of the detector and controls the overall sensitivity that can be achieved, the cell dimensions must be allowed to be as large as possible and account for the major proportion of acceptable band dispersion. If a 4.5% increase in band width is allowed to occur in the detector cell, and 0.5% increase in band width is allowed to occur in the connecting tubes, then two equations can be obtained giving the dimensions of both the detector cell and the connecting tubes.

$$\text{For the detector cell} \quad r = \left(\frac{2.21\, D_M V_R^2}{\pi n Q l}\right)^{1/4}$$

where r and l are the radius and length of the detecting cell respectively and the other symbols have the meaning previously ascribed to them. For the connecting tubes

$$r = \left(\frac{0.24\, D_m V_r^2}{\pi n Q l}\right)^{\frac{1}{4}}$$

where r and l are the radius and length of the connecting tube respectively.

The dimensions of a satisfactory detecting cell and connecting tubes can now be calculated for a given column and given detector combination. The following are very common column and detector characteristics.

Column Length 25 cm , I.D. 4.6 mm , V_0 (Dead Volume) 3 ml,
Flow Rate 1 ml/min = 0.0167 ml /sec,
$D_m = 10^{-5}$ cm^2/sec.

The results obtained using this data in the above equation are shown in table I taking column efficiency values of 4,000; 8,000; 12,000; 16,000; and 20,000 theoretical places. As the efficiency decreases as the retention volume of the solute increases, V_r will be taken as both the value of V_0 (3 ml) the dead volume which will have the maximum efficiency for a given column and at k'=2 where V_r will be 9 ml. At k'=2 the position of a peak in the chromatogram is that which provides the highest resolution in the minimum time. The detector will be taken to have connecting tubes 5 cm long and the detecting cell will be assumed to have a path length of 1 cm. The values of the radius of the cell and connecting tubes together with their respective volumes are given in Table I.

It is seen from Table I that for efficiencies ranging from 4,000 to 20,000 theoretical plates, the maximum cell volume ranges from 3.06 microliters down to 1.37 microliters respectively for a dead volume peak and from 9.18 down to 4.11 microliters respectively for a peak eluted at k'=2. From the results it would appear that the efficiencies of present day microparticulate columns can never be realized for a dead volume peak due to the relatively large detector volumes associated with available detectors. The detector dispersion becomes so significant that efficiency values obtained from them will be much lower than that actually provided by the column. As the equations show, the length of the cell and its radius are interrelated, a cell with a long path length but small radius may meet the same column requirements in an identical manner to an alternative cell of short path length and relatively large radius. In Table II a range of cell diameters

Table I

Dimensions of Detector Cell and Connecting Tube for a Column of Defined Geometry and Different Efficiencies

Efficiency n	Peak Eluted at Dead Volume (3 ml)				Peak Eluted at $k'=2$ (V = 9 ml)			
	Cell I.D.	Cell Volume	Tube I.D.	Tube Volume	Cell I.D.	Cell Volume	Tube I.D.	Tube Volume
4,000	0.624 mm 0.025 in	3.1 µl	0.240 mm 0.010 in	2.3 µl	1.08 mm 0.043 in	9.2 µl	0.416 mm 0.076 in	6.8 µl
8,000	0.525 mm 0.021 in	2.2 µl	0.202 mm 0.008 in	1.6 µl	0.909 mm 0.036 in	6.5 µl	0.349 mm 0.014 in	4.8 µl
12,000	0.474 mm 0.019 in	1.8 µl	0.182 mm 0.007 in	1.3 µl	0.821 mm 0.032 in	5.3 µl	0.316 mm 0.012 in	3.9 µl
16,000	0.441 mm 0.017 in	1.5 µl	0.170 mm 0.007 in	1.1 µl	0.764 mm 0.030 in	4.6 µl	0.294 mm 0.012 in	3.4 µl
20,000	0.417 mm 0.016 in	1.4 µl	0.161 mm 0.006 in	1.0 µl	0.727 mm 0.029 in	4.1 µl	0.278 mm 0.011 in	3.0 µl

Column Length 25 cm
Column Diameter 4.6 mm
Diffusivity of Solute in the Mobile Phase 10^{-5} cm^2/sec
Flow Rate 1 ml/min
Path Length of Cell 1 cm
Length of Connecting Tube 5 cm.

and cell lengths that will meet the requirements for minimum dispersion for a given column system are given. It is seen that at one extreme a cell length 0.25 cm, I.D. 0.59 mm and a volume of 0.68 µl can be used and at the other extreme a cell of length 2.5 cm, I.D. 0.33 mm and a volume of 2.167 microliters could meet the same column requirements. The choice of the best combination of path lengths and cell I.D. that will meet both the minimum dispersion requirements of the column and at the same time provide the maximum sensitivity is somewhat complex. It will depend on the intensity of the light source, the characteristics of the photo cell employed and on the detector electronics. Increasing the path length will increase the response of the detector to a given solute concentration as predicted by Beer's Law. However, as the I.D. of the cell must also be reduced as shown in Table II, this results in a decrease in light intensity falling on the photo cell and thus increases the noise. It follows that the effect on detector sensitivity which has already been shown to depend on the signal to noise ratio may be increased or decreased. The choice of the optimum cell length and I.D. for maximum sensitivity has to be left to the manufacturer who must carefully select and design the light source, optics, photo cell and electronics such that the maximum sensitivity can be realized while employing cell geometries that provide the minimum chromatographic band

Table II
Cell Dimensions that will Provide Minimum Dispersion for a Given Column System

Column System

Column Length 25 cm
Column Diameter 4.6 mm
Dead Volume 3 ml
Flow Rate 1 ml/sec

Diffusivity of Solute in Mobile Phase 10^{-5} cm^2/sec

Cell Length (cm)	Cell I.D. (mm)	Cell Volume μl
0.25	0.59	0.68
0.50	0.50	0.97
1.00	0.42	1.37
1.50	0.38	1.68
2.00	0.35	1.94
2.50	0.33	2.16

dispersion demanded by the column.

Dispersion effects between column and detector can be extremely important where post column reactors are employed. In such systems a reagent is mixed with the column eluent that reacts with the eluted solute to render it detectable by the detecting system employed. Under such conditions band dispersion in the reactor volume between column and detector can be very serious. The design of column reactors will be discussed in the next chapter.

Turbulence

Turbulence in either the detector connecting tube or the cell itself can significantly increase the diffusivity of the solute by convective mixing and reduce band dispersion. However, turbulence in the detector cell itself also results in serious noise and thus reduces the detector sensitivity. It follows that turbulence in the detector should be avoided. Turbulence can be employed in the detector connecting tube to reduce dispersion by interfering with the regular geometry of the tube. Providing the lamina nature of the flow patterns in the tube are broken, this will result in convective mixing and reduce dispersion. This can be achieved by crushing the tube almost flat every 2 mm along its length between the column and detector. However, such a system easily becomes blocked, provides significant back pressure on the column and for these reasons an open tube of the correct dimensions is to be preferred.

Factors that Indirectly Affect Band Dispersion

The band width as drawn by the recorder on the chart paper may be significantly broader than that actually sensed by the detector due to a spreading effect that results from the time constant of the detector amplifier and the recorder itself. Amplifiers and recorders have an inherent time constant that arises from their respective circuit design but a further time constant is often purposely introduced to remove high frequency noise. If this time constant is of commensurate period to the time standard deviation of the eluted peak then the peak will suffer significant broadening. The effect of amplifier time constant on the shape of detected chromatographic peaks has been elegantly treated by Vandenheuvel (3), Schmauch (4), and Sternberg (5). For those readers wishing to study the effect of time constants on peak distortion in detectors they are recommended to read the work of Sternberg.

Sternberg developed the following equation to describe the peak shape after being distorted by an amplifier with a time constant of T'.

$$X = X_o e^{-t/T'} /T' \int e^{t/T} e^{-(t-t_o)/2T_t^2}$$

where X = the voltage output of the amplifier to the recorder
X_o = a constant
t_o = time at the peak maximum
T_t = the time standard deviation of the eluted peak
t = the elapsed time

The explicit solution of the above equation is rather complicated, but the distortion of a normal Gaussian peak can be determined from the above equation by the use of a computer. Consider a column 25 cm long and 4.6 mm in diameter having a dead volume (V_o) of 3 ml, an efficiency (n) of 12,000 theoretical plates and operated at a flow rate of 1 ml/min (Q). From the plate theory the standard deviation of the dead volume peak in ml of mobile phase will be

$$\frac{V_o}{n^{\frac{1}{2}}} = \frac{3}{(12,000)^{\frac{1}{2}}} = 0.0274 \text{ ml}$$

Thus the time standard deviation

$$T_t = \frac{0.0274}{Q} = 0.0274 \times 60 = 1.64 \text{ sec.}$$

Thus taking values for the time constant of the amplifier of 0.6 and 1.5 secs and replacing T_t in the equation by the calculated value of 1.64 sec,

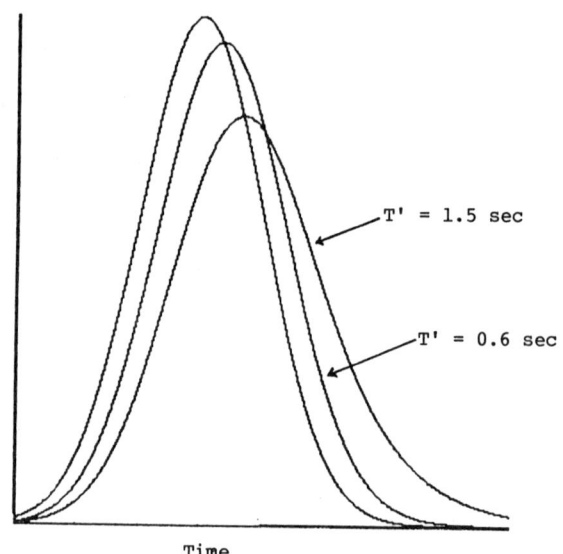

Figure 1   Curves Demonstrating Peak Distortion Resulting from Significant Amplifier Time Constant

the shape of the resultant peaks can be calculated. Using a computer the resultant curves were determined over the range of t = 4.92 sec to t = 8.2 sec and the results obtained are shown in figure 1 together with the original undistorted peak. It is seen that the effect of a 1.5 second time constant is to seriously distort the peak both by increasing the peak width and reducing the peak height. This has resulted in a reduction of the resolving power of the column and the sensitivity of the detector. It is,

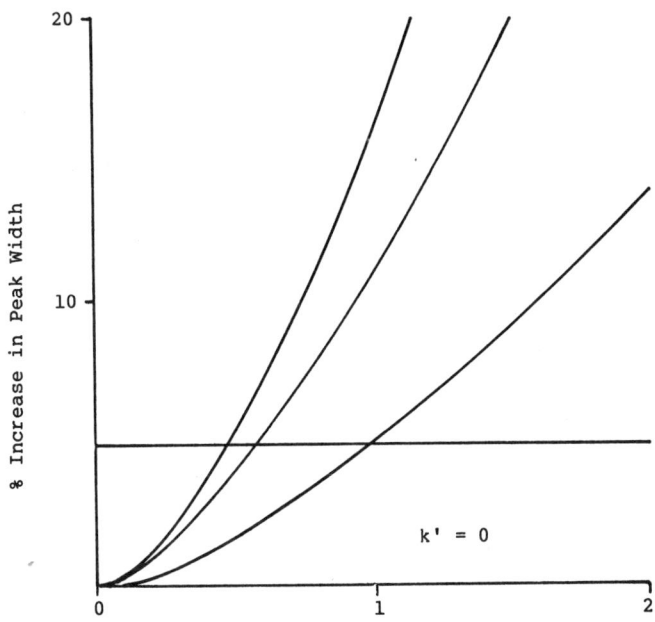

Time Constant in sec

Column Length 25 cm   Column Diameter 4.6 mm I. D.

Dead Volume 3 ml

Figure 2   Curves Relating the Percentage Increase in Peak Width to the Amplifier Time Constant

therefore, important to determine the maximum time constant that can be employed with any given column. Using the same equation and again with the aid of a computer the percentage increase in peak width that will occur for a range of different time constants can be calculated. The results obtained for the dead volume peak from columns of the same dimensions as given previously but with efficiencies of 4,000; 12,000; and 20,000 theoretical plates are given in figure 2. If an increase in band width of 5% is tolerated which will result in a reduction in peak height of about 5%, then the maximum permitted amplifier time constant will be 0.48 sec, 0.58 sec and 1 sec for peaks having time standard deviations of 1.26 sec, 1.64 sec and 2.85 sec respectively. Thus the maximum time constant that can be tolerated is seen to be consistently about 35% of the time standard deviation of the peak. In figure 3 the same calculations have been carried out to provide similar curves but for a solute eluted at k' = 2. It is seen that the maximum time constants that can be tolerated are now 1.32 sec, 1.76 sec and 3.04 sec for peaks having standard deviations of 3.82 sec, 4.93 sec and 8.54 sec respectively. It should also be noted that the maximum amplifier time constant is again consistently 35% of the time standard deviation of the peak.

The problem can be approached in an entirely different manner by using the principle of the summation of variances. The effect of amplifier time constant is to combine two functions, an exponential function and a Gaussian function. As these two functions describe physical phenomena that are not interacting in the sense that they proceed independently of one another, the variance of the combined function is equivalent to the sum of the variances of each individual function. The time variances of an exponential function of the form

$$e^{-\frac{t}{T_1}} \text{ is } T_1^2$$

and that of a Gaussian function

$$e^{\frac{-t^2}{2T_t^2}} \text{ is } T_t^2$$

It follows that

$$T_1^2 + T_t^2 = T^2$$

where T^2 is the time variance of the resulting peak as described by the recorder. Now T_t^2 is the time variance of the solute band leaving the column and thus if a band width increase of 5% is considered acceptable ($T = 1.05\ T_t$)

then $T_t^2 + T_1^2 = (1.05\ T_t)^2 = 1.103\ T_t^2$

thus $T_1^2 = 0.103\ T_t^2$

and $T_1 = 0.32\ T_t$

Thus the maximum value of T_1 that can be tolerated to maintain column resolution will be 32% of the time standard deviation of the eluted peak. The discrepancy between the value of 32% derived in this manner to the value of 35% derived from computer integration results partly from the fact that in digital integration finite and not infinitesimal steps are used during

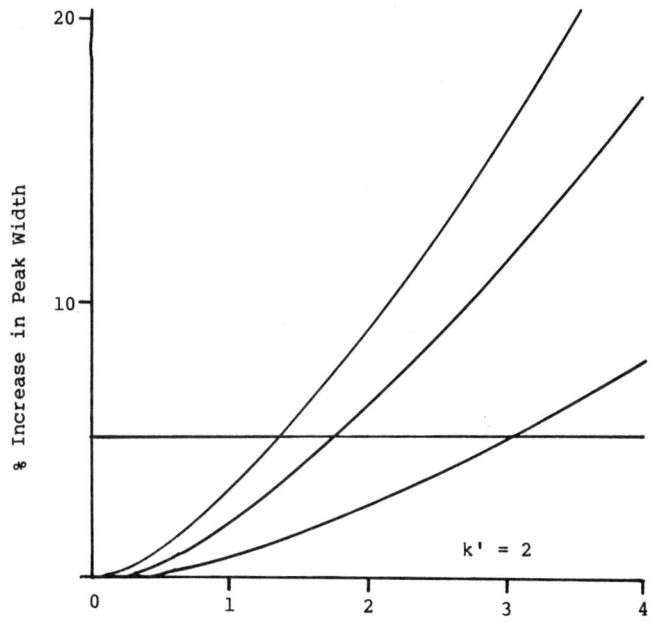

Time Constant in sec

Column Length 25 cm, Column Diameter 4.6 mm I. D.

Dead Volume 3 ml

Figure 3   Curve Relating the Percentage Increase in Peak Width to the Amplifier Time Constant

integration procedure and partly due to the assumption that the variance of the distorted peak is at 0.607 of the peak height. The more correct value is, therefore, given by the second treatment, namely, 32%. It follows that for any column the maximum permitted time constant T' will be given by

$$T' = \frac{0.32\, V_R}{n^{\frac{1}{2}} Q}$$

Where V_R and n are the retention volume and efficiency of the peak and Q is the flow rate through the column. V_R should normally be taken for the dead

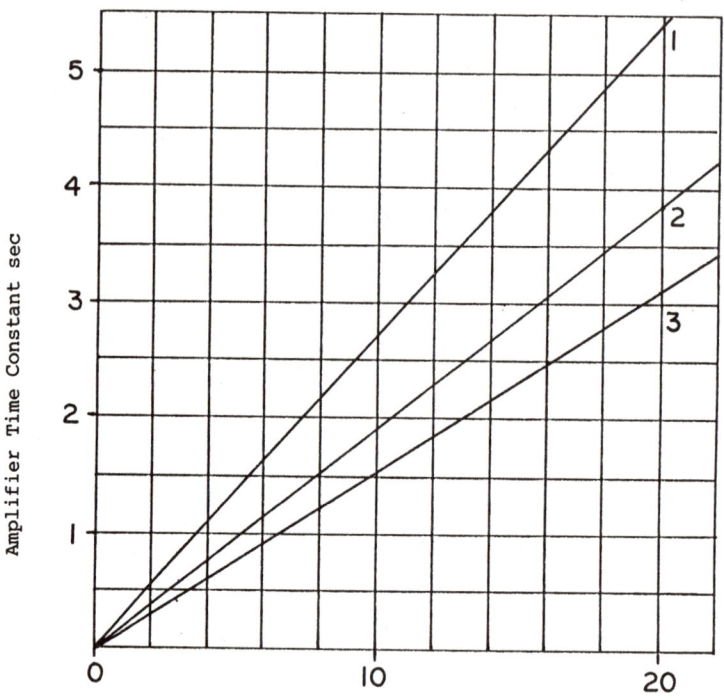

1. Column Efficiency 5,000 Theoretical Plates
2. Column Efficiency 10,000 Theoretical Plates
3. Column Efficiency 15,000 Theoretical Plates

Figure 4 Curves Relating Retention Volume to Amplifier Time Constant for Columns of Different Efficiencies

volume peak as it has the smallest retention volume and the highest efficiency. The above equation for T is applicable to columns of all dimensions packed with any stationary phase and employed with any mobile phase. In figure 4 curves are shown relating the minimum amplifier time constant that can be permitted to maintain column resolution, to solute retention volume. The individual curves are for columns of different efficiency operated at a flow rate of 1 ml/min. The appropriate time constant for any column system, within the range given, can be obtained by interpolation.

The Time Constant of the Recorder

The potentiometric recorder does not have a time constant of the form normally associated with the amplifier which, in general, results from a capacity resistance network intrinsic in the amplifier circuit. The response of an amplifier to an instantaneous applied constant voltage is normally an exponential function of time, whereas for a potentiometric recorder the response is usually linearly related to time. The linear response results from the feed back circuitry incorporated in the sensor system which is necessary for stability. In figure 5 the recorded reading is plotted against time for an instantaneous applied constant 9 mv signal. The recorder was the Honeywell Electronik 196, 10 mv potentiometric recorder operated with a chart speed of 1 cm/sec and having a specified balancing time of 0.5 sec. It is seen that the response is approximately linear.

Now for a linear function the time variance T^2, is given by

$$T^2 = \frac{t_R^2}{18}$$

where t_R is the time taken for the recorder to reach the applied voltage. In the example given

$$T^2 = \frac{(0.79)^2}{18} = 0.035 \text{ sec}^2$$

and thus the time standard deviation, $T = 0.19$ sec.

It is seen that if a recorder is employed having a balancing time of about 0.5 sec the contribution in time variance to the eluted peak is relatively small and by itself would not significantly increase the width of the eluted peak and reduce resolution. However, it is yet another contribution to band dispersion and it is the sum of all the dispersive effects that take place subsequent to the column that ultimately limits the chromatographer from realizing the actual efficiency that the column provides.

Figure 5

Response Curve of a Potentiometric Recorder

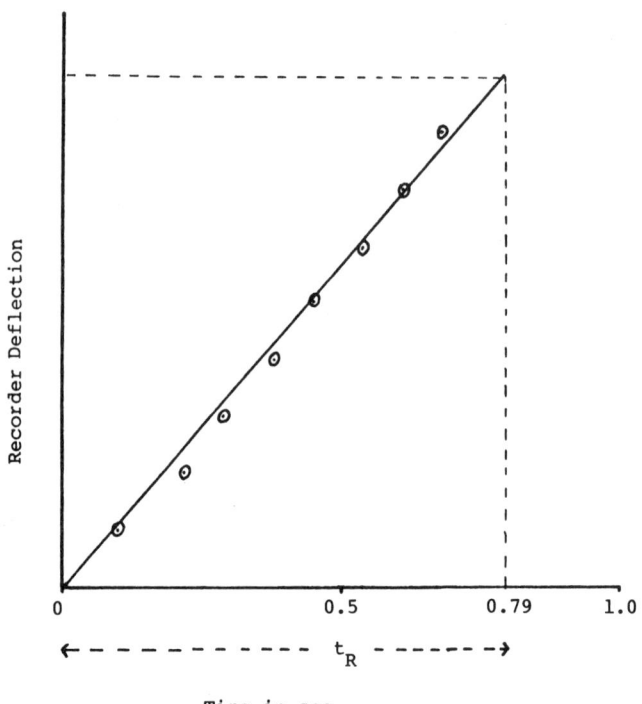

Adsorption Effects in the Detector

When using a mobile phase that contains a small percentage of a polar solvent contained in a nonpolar solvent, for example ether or ethanol in heptane, a layer of the more polar solvent is often adsorbed on the surface of the quartz or glass detector cell. If a solute is eluted that is more polar than the ether or ethanol then the polar solvent is displaced from the cell walls and is replaced by the solute. This displacement results in a spurious peak or, more often, a distorted peak shape. The effect of adsorption on the walls of the detecting cell cannot be treated quantitatively and when it occurs it should be eliminated by choosing an alternative mobile phase.

Summary

There are a number of band dispersing processes that occur in various parts of the detector and its associated electronics. By careful design, any one of these processes can usually be reduced to a level where it does not significantly affect column performance. However the total dispersion that takes place is a combination of all these effects and thus the resulting time variance of an eluted peak T^2 will be given by

$$T^2 = T_C^2 + T_1^2 + T_2^2 + T_3^2 + T_4^2$$

where

T_C^2 is the time variance of the peak leaving the column.

T_1^2 is the time variance resulting from dispersion in the detector connecting tubes.

T_2^2 is the time variance resulting from dispersion in the detector cell.

T_3^2 is the time variance resulting from the time constant of the amplifier.

T_4^2 is the time variance resulting from the time constant of the recorder.

If a 5% increase in band width is tolerated then

$$T^2 = (1.05 \text{ C})^{\frac{1}{2}}$$

$$T_1^2 + T_2^2 + T_3^2 + T_4^2 = 0.103 \; T_C^2$$

and

$$(T_1^2 + T_2^2 + T_3^2 + T_4^2)^{\frac{1}{2}} = 0.32 \; T_C.$$

If it is assumed that in the future a 25 cm long column 4.6 mm I.D. with a dead volume of 3 ml and operated at a flow rate of 1 ml/min will provide 20,000 theoretical plates then the time standard duration of the peak, T_c, will be about 1.27 sec.

It follows that

$$(T_1^2 + T_2^2 + T_3^2 + T_4^2)^{\frac{1}{2}} = 0.41 \text{ sec.}$$

It is seen from the above equation that the detecting system has to be very carefully designed if the efficiency of 20,000 theoretical plates is to be realized. It is an unfortunate fact that with many present day detecting systems for liquid chromatography, an efficiency of even 10,000 theoretical plates from such a column would never be seen. If 10,000 theoretical plates were indeed achieved then the column would, in fact, be providing very much more than the 10,000 theoretical plates measured. It is hoped that the detector manufacturers will, in the future, design detectors having specifications that are appropriate to the column efficiencies presently obtainable and also be suitable for even higher efficiencies that will result from future developments in column technology.

REFERENCES

1. A. Klinkenberg, Gas Chromatography 1960, (Ed. R. P. W. Scott), Butterworth, London, 1960, p. 182.
2. R. P. W. Scott and P. Kucera, J. Chromatogr. Sci., 9 (1971) 641.
3. F. A. Vandenheuvel, Anal. Chem., 35 (1963) 1193.
4. L. S. Schmauch, Anal. Chem., 31 (1959) 225.
5. J. C. Sternberg, Advances in Chromatography, (Ed. J. C. Giddings and R. Keller), Marcel Dekker, New York, Vol. II, 1966, p. 206.

CHAPTER 4

Summary of Detector Criteria

The pertinent properties of a liquid chromatography detector that can influence the efficiency of any chromatographic system with which it is associated have been discussed in detail in the previous chapters. A list of specifications can now be drawn up and their pertinence to the effective operation of the liquid chromatograph summarized.

1) The Dynamic Range

The dynamic range gives the concentration range over which the detector will give a response and has been given the symbol D_R. The range should be given in units of g/ml for concentration sensitive detectors and g/sec for detectors sensitive to the mass passing through them per unit time. The value of D_R is particularly useful when the detector is being used for preparative separations.

2) The Response Index

The response index of a detector is a measure of its linearity and has been given the symbol r. It is a dimensionless constant and for a truly linear detector should take the value of unity. In practice the value of r should be between 0.98-1.02 and is very important where accurate quantitative analyses are required.

3) The Linear Dynamic Range

The linear dynamic range gives the concentration range over which the detector's response is linear within the tolerance given by the response index and is given in units of g/ml for a concentration sensitive detector and in g/sec for mass sensitive detectors. It has been given the symbol D_L and in conjunction with the response index is important where accurate qualitative analysis is required. A knowledge of the linear dynamic range of a detector is particularly valuable where mixtures containing components present at widely different percentage levels are being analyzed.

4) **The Detector Response**

The detector response has been given the symbol R_c or R_m depending on whether the detector is concentration sensitive or mass sensitive and is given in units of mv/g/ml and mv/g/sec respectively. In conjunction with the noise level of the detector it permits the sensitivity of the detector to be calculated and also aids in comparing the performance of detectors of the same type.

5) **Noise Level**

The noise level of a detector is measured in mv and is taken as the maximum amplitude of the combined short and long term noise taken over a period of about ten minutes. It has been given the symbol N_D and is used in determining the detector sensitivity and in comparing the stability of different detectors.

6) **Detector Sensitivity**

The detector sensitivity is the concentration of solute, or the mass entering the detector per unit time that will provide a signal to noise ratio of two. It has been given the symbols S_c and S_m and is quoted in units of g/ml and g/sec respectively. A knowledge of the detector sensitivity is required in practice for trace analysis and to assess the compatability of the detector with the loading capacity of the column with which it is to be associated. It is also necessary to determine the magnitude of the charge required for a given separation and to calculate the maximum peak capacity that can be realized from the column employed.

7) **Connecting Tube Dimensions**

The length, internal diameter and volume of the connecting tube (l_r, d_r and v_r) are given in the units of cm, mm and μl respectively and are used with the cell dimensions to determine the maximum efficiency that can be realized from any associated column.

8) Cell Dimensions

The cell volume is given in microliters and has the symbol v_c. The length and internal diameter of the cell (l_c, d_c) are given in cm and mm respectively. The values of l_c and d_c together with dimensions of the connecting tube condition the maximum efficiency that can be realized from any column associated with it.

9) The Amplifier Time Constant

The amplifier time constant is given in sec and has the symbol T. The value of T also conditions the maximum efficiency that can be realized from any column associated with the detector.

Table I

Detector Criteria Pertinent to Liquid Chromatography

Specification	Units	Symbol
Dynamic Range	g/ml g/sec	D_R
Response Index	–	r
Linear Dynamic Range	g/ml g/sec	D_L
Detector Response	mv/g/ml mv/g/sec	R_c R_m
Noise Level	mv	N_D
Detector Sensitivity	g/ml g/sec	S_c S_m
Cell Dimension		
Length	cm	l_c
Diameter	mm	d_c
Volume	µl	v_c
Connecting Tube Dimension		
Length	cm	l_t
Diameter	mm	d_t
Volume	µl	v_t
Amplifier Time Constant	sec	T

The pertinent data is given in a concise form in Table I. The data is essential for the effective use of a detector in liquid chromatographic analysis. Unfortunately at present very few manufacturers provide such a comprehensive set of data with their detectors and it may be necessary for these to be determined by the chromatographer. However it is hoped that in the future the detector manufacturers will appreciate the need for such data and optimize their present designs to be appropriate to the requirements of the modern high efficiency columns presently available.

In dealing with band dispersion that takes place in the detector sensing cell it was assumed that the cell took the form of a cylinder. In many commercial instruments, however, the sensing cell is not cylindrical but has a square, triangular or rectangular cross section. The effect of such cross section geometry on band dispersion is somewhat complex and in general there is significantly greater dispersion in a tube of triangular or rectangular cross section than in a tube of equivalent length and volume but cylindrical in form. As an approximation the radius of a cylindrical cell giving the equivalent dispersion can be taken as 1.5 times the radius of a tube that has the same cross sectional area as the triangular, rectangular or square cross section form.

Chapter 5

Ancillary Equipment

Predetector Reactors

In classical liquid-liquid and liquid-solid chromatography, the technique of reacting sample components to form products more responsive to the detection system is well established. Derivatization or complexation may be carried out on the bulk sample before chromatography, or on the eluted components as they emerge from the chromatographic column. The former procedure is more desirable but is dependent on the derivatives being stable and amenable to separation; the latter is more common and more generally useful because established separation procedures can be used as in the case of the ninhydrin reagent and more recently the fluorescent reagents for the detection of amino acids and peptides. Derivatization of components after chromatography as they emerge from the column necessitates the introduction of some form of continuous reactor which, inevitably, leads to some broadening of the chromatographic bands. The specification of such reactors therefore needs to be determined particularly as the permitted dispersion, when using high efficiency columns, can be very small.

In practice the eluent from the column is mixed with the reagent in a flow-through mixing chamber situated between the column and the detector and the reaction mixture passes into the detector in the normal manner. The system should be designed to meet certain criteria.

1. The band dispersion resulting from the mixing of the eluent with the reagent shall not exceed 5% of the dispersion of the band as it leaves the column in order to maintain the resolving power of the column.

2. Mixing between the column eluent and the reagent should be complete to ensure efficient reaction.

3. The dwell time of the mixture in the mixing vessel should be sufficient to ensure complete reaction.

Now the criteria given in (3) will be governed by the criteria given in (1) so that the conditions of reaction such as pH, reagent concentration, etc. may have to be adjusted to provide an adequate reaction rate to meet the requirements of (1) and (3).

Assuming that the reactor is designed to meet the requirement given in (2) then the mixing chamber becomes a dilution vessel and if t is the elapsed time

$$X = X_o \exp(-Qt/V) = X_o \exp(-t/T_R)$$

where X is the concentration of reacted solutes at time t in g/ml

X_o is the initial concentration of reacted solute in g/ml

Q is the combined flow rate of the column eluent and the reagent in ml/sec

V is the volume of the mixing vessel in ml

and $T_R = V/Q$ the time constant of the dilution system in sec

Now from the work of Sternberg (1) a similar treatment can be used as in the case of the effect of the time constant of the detector amplifier.

$$T_C^2 + T_R^2 = T^2$$

where T_C^2 is the time variance of the peak leaving the column

T_R^2 is the time variance of the mixing vessel

T^2 is the time variance of the peak leaving the reactor

Now assuming $T = 1.05\ T_C$

that is a 5% increase in band width is tolerated

$$T_C^2 + T_R^2 = (1.05\ T_C)^2 = 1.103\ T_C^2$$

thus $T_R^2 = 0.103\ T_C^2$

and $T_R = 0.32\ T_C$

thus $V = 0.32\ QT_C$ \hfill (1)

Further the mean reaction time for the solute and reagent will be

$$V/Q = 0.32\ T_C \hfill (2)$$

From equations 1 and 2 the volume of the reactor and the reaction time can be calculated.

In table I the values for T_C, V and reaction time V/Q are given for a 25 cm column 4.6 mm I.D. having a dead volume of 3 ml and efficiencies of 6,000, 12,000 and 18,000 theoretical plates. Values are given for solutes eluted at the dead volume and at a k' value of 2. The column flow rate and reactor flow rate were both taken as 1 ml/min (i.e. 0.0167 ml/sec) giving a total flow rate through the reactor of 2 ml/min (i.e. 0.0334 ml/sec). It is seen that both reactor volume and reaction times are extremely small. At one extreme, for a column of 4000 theoretical plates and for a solute eluted

Table I

Mixing Vessel Volumes and Reaction Times for a Reactor used with a Standard Column having Different Efficiencies

Column Efficiency (Theoretical Plates)	T_c	k' = 0 Reactor Volume	Reaction Time	T_c	k' = 2 Reactor Volume	Reaction Time
6,000	2.32 sec	24.8 µl	0.74 sec	6.96	74.4 µl	2.23 sec
12,000	1.64 sec	17.5 µl	0.53 sec	4.92	52.5 µl	1.59 sec
18,000	1.34 sec	14.3 µl	0.43 sec	4.02	42.9 µl	1.29 sec

at a k' value of 2, the reactor volume must be less than 75 µl and the reaction must be complete in 2.3 sec. The conditions are even more stringent for a solute eluted close to the dead volume from a column of 18,000 theoretical plates where the reactor volume cannot be more than 15 µl and the reaction must be complete in 1.3 sec. It follows that when using high efficiency microparticulate columns care must be taken to develop the reaction conditions so that reaction is complete in the necessary limited dwell time available in the reactor.

A simple design of reactor is shown in figure 1. It consists of a 1/16 in T piece that has been drilled out to allow the 1/16 in O.D., 0.010 in I.D. tubing entering each arm of the T to meet at the center. The flow from the column and to the detector is directly in line and the required reactor volume achieved by adjusting the distance between the ends of the two tubes. The reagent is arranged to enter the chamber via the other limb which should be in contact with the end of the tube carrying the column eluent into the reactor as shown in figure 1. Mixing is achieved by the turbulence resulting when the reagent and column eluent streams meet. This reactor design is quite adequate for flow rates of 0.5 ml/min or more but may not be so satisfactory at lower flow rates where there will be less turbulence and possibly incomplete mixing. It should be emphasized that the effect of the

Figure 1

A Simple Post Column Reactor

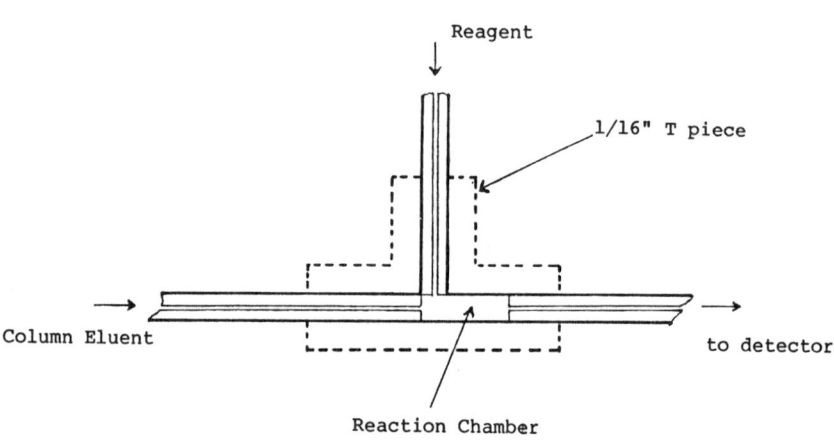

reactor volume on band dispersion is still not generally recognized and workers using such systems should be cautious in the design of their reactor when using them in conjunction with high efficiency columns.

An alternative approach to reactor design was discussed by Snyder (2) where segmented flow is employed to minimize dispersion in the reactor and in the connecting tube to the detector. Segmented flow is achieved by regularly introducing a bubble of a permanent inert gas into the column eluent subsequent to it being mixed with the appropriate reagent. The gas bubbles break up the parabolic velocity profile of the liquid that normally occurs in a tube and which is the major cause of band dispersion. Snyder gives a detailed mathematical treatment of segmented flow and devises equations that give the design and operating conditions of the reactor to provide minimized dispersion. Segmented flow techniques render the system more cumbersome particularly as the gas bubbles have to be removed from the eluent prior to detection, however, the system is particularly appropriate

Noise Filters

Noise filters are employed to reduce short term noise in the detector output and thus enhance the signal to noise ratio which has been shown in Part 1 Chapter 2 to also increase the overall sensitivity of the detector. Unfortunately, however, it is not the short term noise that usually restricts the detector sensitivity but long term noise which is of the same order of frequency as the eluted peak itself. Further if a noise filter is used to reduce long term noise then this must result in reduction of peak height and an increase in band width which will affect the resolving power of the column. Nevertheless there are frequent times when the use of a noise filter is advantageous and so their characteristics will be discussed and their effect on band dispersion considered.

There are two types of noise filter, the passive filter and the active filter, the passive filter consists simply of a resistance capacity network having a time constant suitable for the noise to be filtered. The active filter is a similar network but used in conjunction with an operational amplifier. The basic difference between the two types of filter is sharpness of frequency cut off. This can be best explained by reference to figure 2 where the percentage of the signal passed by the filter is shown plotted against frequency. The continuous curve is for the passive filter and the broken curve for the active filter. Consider the situation where the peak has a frequency of f_1 and it is required to remove noise having a frequency of f_2. Now it is seen from figure 2 that using the passive filter 60% of the noise at frequency f_2 is removed but at the expense of 25% of the signal at frequency of f_1. However with the active filter the noise has been reduced to 20% of its original value while the signal has not been reduced at all. Thus with the passive filter the signal to noise ratio has been increased only by a factor of 1.9 whereas the active filter has increased the signal to noise ratio by 5. The advantage of the active filter results from the relative slopes of the two curves and it is seen that using the active filter the frequency of the rejected noise can be made to be much closer to the frequency of the signal without affecting the magnitude of the signal.

An example of a commercially available active filter suitable for use in liquid chromatography is that manufactured by Spectrum Scientific Corp.

Figure 2

Transmission Curves for Active and Passive Filters

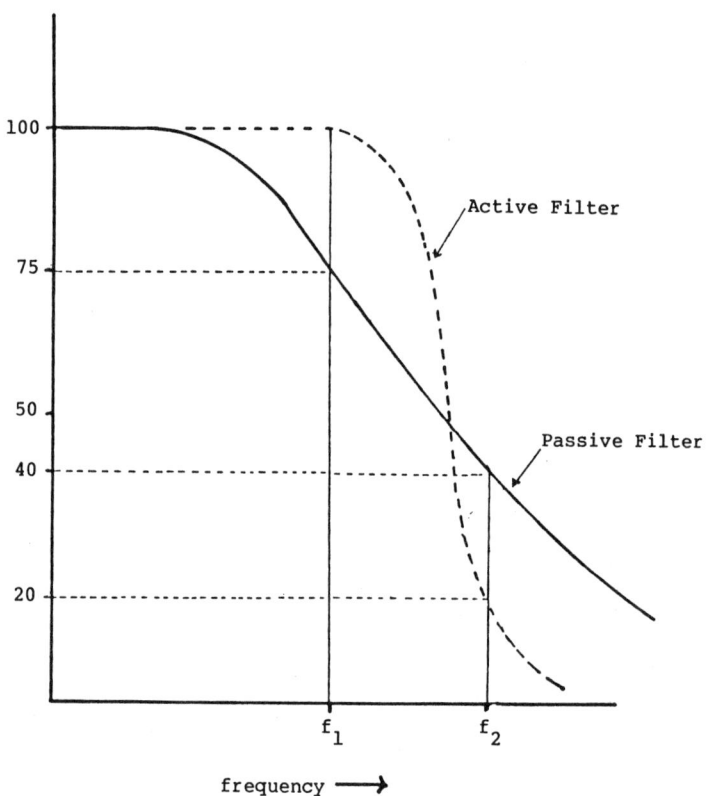

This filter also has amplification capabilites of 1, 2, 5 and 10. The cut off frequencies available on the instrument are 0.01, 0.02, 0.05, 0.1, 0.5, and 1 Hz which, although they include the range of noise frequencies met within chromatography are not ideally suited for chromatographic application. An example of an application of this filter is shown in figure 3. The example is actually from a gas chromatograph but serves to illustrate the dramatic effect of an active filter on noise elimination; the small peak on the side of the major peak is completely lost in the unfiltered chromatogram. However the introduction of a noise filter is in

Figure 3

An Example of Use of the Active Filter in Chromatography

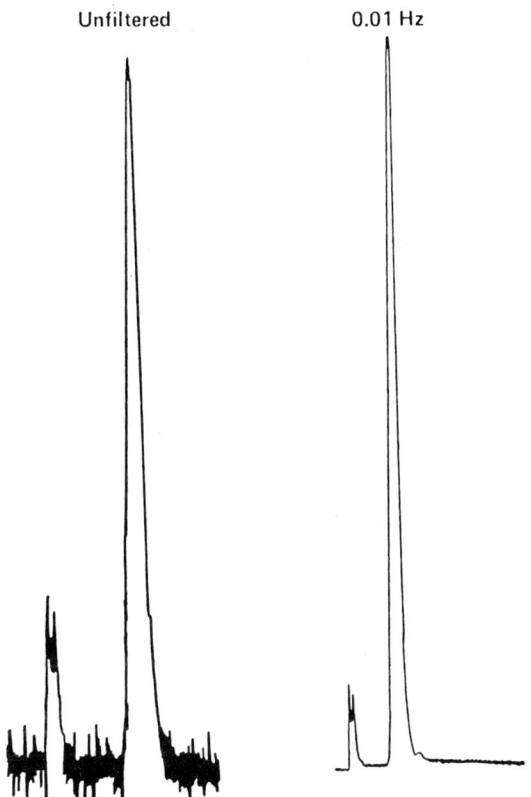

effect equivalent to introducing a time constant into the detector electronics and will thus produce band dispersion and, as has already been shown, the time constant must not exceed 32% of the time standard deviation of the eluted peak.

In Table II the time constant experimentally determined for the first 5 cut-off frequency settings of the Spectrum filter are shown together with the minimum time standard deviation of the chromatographic peaks that can be used with them. In the last column of Table II the maximum column efficiency is given (calculated for the dead volume peak) that can be used

Table II

Operating Data for the Spectrum Active Filter

Cut-off Frequency	Time Constant	Minimum T_r for eluted peak	Maximum Usable Column Efficiency
0.01 Hz	5.7 sec	17.8 sec	102
0.02 Hz	2.4 sec	7.5 sec	576
0.05 Hz	1.3 sec	4.1 sec	1,980
0.10 Hz	0.55 sec	1.7 sec	11,200
0.50 Hz	0.20 sec	0.63 sec	81,600

T_r is the time standard deviation of the eluted peak.

Column 25 cm long, 4.6 mm I.D., dead volume 3 ml operated at a flow rate of 1 ml/min.

for each frequency setting of the filter. The data is for a column 25 cm long, 4.6 mm I.D. having a dead volume of 3 ml and operated at a flow rate of 1 ml/min. The great advantage in using an active filter as opposed to a passive filter is clearly seen. Even when a cut-off frequency of 0.05 Hz (which is equivalent to a period of 20 sec) is employed the effective time constant from the point of view of band dispersion is still only 1.30 sec which permits columns with efficiencies of up to about 2000 plates to be employed. For optimum filtering the minimum frequency should be chosen that is acceptable with respect to the column efficiency. For example, for a column of 2000 theoretical plates it is seen from the table that frequencies of 0.5, 0.1 or 0.05 Hz could be employed but the minimum value 0.05 Hz should be used to provide the maximum increase in the signal to noise ratio and thus the highest sensitivity. The first ranges of filtering are of little value to chromatographers and the filter would be of much greater use if the total range of cut-off frequencies could be between the levels of 0.05 Hz and 0.50 Hz. Six frequency ranges between these limits would allow a more accurate choice of filtering that would be appropriate to columns presently available. The active filter has, potentially, great value in increasing the sensitivity of chromatographic detectors and those readers wishing to learn more about the subject are recommended to read the review article on active filters published in Electronics (3) and a paper entitled "Signal Conditioning in the Journal of Instruments and Control Systems" (4).

Integration

The integration of the detector output with respect to time gives a value for the area under an eluted peak and thus is useful for quantitative analysis. It is not intended here to give an account of data handling procedures in liquid chromatography as this is a subject that is sufficiently extensive to warrant a book to itself. However, the various methods of integration will be described and their various merits discussed.

Electromechanical Methods

The first automatic integrating system to be used in chromatography was based on the inertia motor. The voltage to the inertia motor is normally derived from a secondary slidewire on the recorder. The moving arm of this slidewire is coupled mechanically to the recorder pen carriage which means that a voltage proportional to pen deflection is applied to the motor. The integrating motor rotates a disc that interrupts the light path to a photocell and gives rise to pulses which activate mechanical counters. As the motor is made to rotate at a rate proportional to the pen deflection, it will sum the area under a chromatographic peak. Although this system produces a record in which area count is displayed under each peak, it suffers from the disadvantage that a certain current level has to be exceeded before the motor starts to rotate, this in effect is a "dead band" and results in a slice of area across the base of each peak not being measured.

Another electromechanical system is based on the ball and disc principle. The ball is positioned on the disc (by mechanical linkage) at a distance from the disc center that is proportional to the deflection of the recorder pen from the baseline zero. The disc is rotated at a constant speed and thus the ball rotates at a speed proportional to its distance from the center of the disc which will also be proportional to the pen deflection. The movement of the ball is transmitted by a suitable mechanical system to a disc that interrupts the light falling on a photocell and the pulses from the photocell are then fed to a suitable counter. Mechanical integrators are relatively inexpensive but have been largely superceded by the digital integrator and the computer.

The Analog Electronic Integrator

The analog electronic amplifier is essentially an active filter with an infinite time constant. The signal from the detector is made to charge a condenser and the potential across the condenser resulting from the integral of current with respect to time during the elution of a peak is thus proportional to the area under the peak and, if fed to a suitable recorder, will provide the integral type of curve. Mathematically this can be put in the form

$$V = \int_0^t \frac{idt}{C}$$

where V is the voltage across the condenser
C is the capacity of the condenser
i is the current output of the detector
and t is the elution time of the peak

The integrating electronic amplifier is now seldom used. Its big disadvantage is that the base line must not change during the entire chromatographic development. Even if the base line shift is extremely small, this difference signal will be continuously integrated and produce a sloping base line on the integral chromatogram. The chromatographic conditions must therefore remain absolutely constant and even when this is achieved the adjustment of the electrical zero before a run can be extremely tedious and time consuming.

The Digital Integrator

The digital integrator comprises of a number of circuits, each responsible for a particular function that together provide retention times, peak areas, in the form of a printed statement, and also compensate for base line drift. A block diagram of a digital integrator system is shown in figure 4. The system is simple in form and has been chosen to illustrate the principles involved. Most practical systems presently available are far more complex than that shown in figure 4. The signal from the detector

Figure 4

Diagram of Digital Integrator

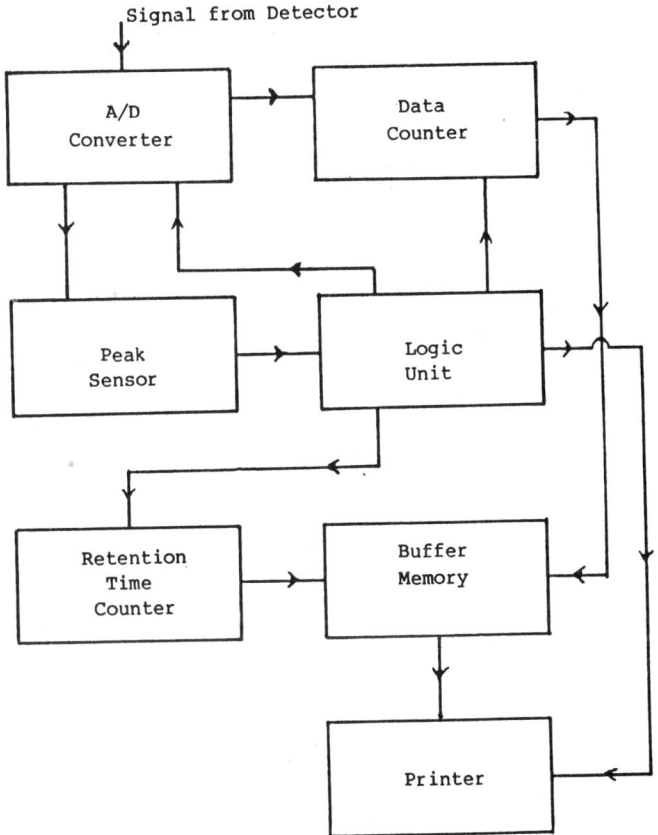

enters an analog to digital converter (A/D) which regularly samples the input and converts the instantaneous sampled voltage to electrical pulses, the number of which is proportional to the voltage sampled at that instant. The output from the A/D converter passes to both a peak sensor and a data counter. The data counter will, on receiving a signal from the logic unit, commence counting the pulses until stopped by the logic unit. The peak sensor continually compares consecutive counts from the A/D converter and if it senses a difference it sends a signal to the logic unit. The peak sensor in fact differentiates the digitized detector signal. On receiving a signal from the peak sensor the logic unit will decide whether the change in detector signal results from drift (a small relatively constant positive or negative differential signal) or from the start of a peak (a relatively large positive increasing differential signal). If the signal change represents drift, the logic unit will control the A/D converter to compensate for the drift. If the signal change indicates the start of a peak the logic unit will initiate the data counter. When subsequently the peak sensor indicates a large increasing negative differential the logic unit will interpret this as a peak maximum and will activate the time counter. The time counter is a digital clock that is initiated at the time of sample injection. On receiving a signal from the logic unit it passes the time elapsed from the injection (that is the retention time of the peak maximum) to be temporarily stored in the buffer memory. When the peak sensor indicates that the differential signal has fallen to zero or when the value has reached a predetermined minimum the logic unit recognizes the end of a peak. It then instructs the data counter to pass its total count from the beginning of the peak (the peak area) to the buffer storage and then resets the data counter to zero ready for the next peak. Subsequently the logic unit instructs the printer to print out the contents of the buffer memory (providing the retention time and peak area) and then clears the buffer memory ready for the next peak. The retention time counter is only set to zero at sample introduction. In this way the retention time and area of each peak are recorded.

In general, digital integration is far more accurate than other methods of integration and can be achieved either by using a digital integrator of the form described above or by means of a computer. The digital integrator has to work in real time in the sense that it provides data as the peaks are eluted and is "hard wired". That is to say, the logic of the system is fixed as the sequence of measurements is controlled by the nature of the electrical components used and such a system is thus inflexible. The computer on the other hand, in its more sophisticated mode, does not operate in real time but stores the data from the complete chromatogram. The data is subsequently processed to produce the chromatographic data by "software". "Soft-ware" is the name for a specific set of instructions in the form of programs and the program or instruction can be changed or modified at any time. Computer data handling thus can be completely flexible and can be made to suit the type of chromatograph employed. Data handling by computer is also far more expensive.

REFERENCES

1. J. C. Sternberg, Advances in Chromatography, Vol. 2, (Ed. J. C. Giddings and R. A. Keller), Marcel Dekker, New York, 1966, p. 205.
2. L. R. Snyder, J. Chromatogr., 125 (1976) 287.
3. L. Mattera, Electronics,   (1972) 104.
4. D. H. Nalle, Instr. Contr. Syst., 42 (1969) 77.

PART 2

BULK PROPERTY DETECTORS

CHAPTER 1

General Characteristics of Bulk Property Detectors

One of the first on-line liquid chromatography detectors to be developed, the Refractive Index Detector, was a bulk property detector. Bulk property detectors continuously monitor some physical property of the column eluent and by the use of a suitable transducer provide a voltage - time output that is made proportional to the physical property being measured. The properties of the mobile phase that are most commonly monitored in commercially available detectors of this type are refractive index, dielectric constant and electrical conductivity. Any solute eluted will be detected providing the magnitude of the physical property of the solute differs sufficiently from that of the mobile phase. The sensing device usually consists of a cylindrical cell of small volume that is either made part of an optical measuring system or is fitted with suitable electrodes to measure some electrical property of the cell contents. In some instances a reference cell is also provided that is filled with pure mobile phase and the output from the measuring cell and reference cell are compared, the difference providing the output to the recorder. Such reference cells are always included in the refractive index detector as they help to compensate for changes in the light intensity emitted from the light source and to some extent changes in ambient temperature. In effect, reference cells help to reduce the noise level of the detector and thus increase its sensitivity.

The Limitations of Bulk Property Detectors

Bulk property detectors have a very restricted sensitivity that is directly due to the principle on which they function. Consider a bulk property detector that monitors the density of the eluent leaving the column and assume that it is required to detect a concentration of a dense material, such as carbon tetrachloride (S.G. 1.595), at a level of one microgram per ml in heptane (S.G. 0.684). This situation will be favorable for such a detector as the solute to be detected exhibits a large difference in density from that of the mobile phase.

Let the change in density resulting from the presence of the solute at a concentration of 10^{-6} g/ml be Δd.

It follows that $\Delta d = (d_1 - d_2) X_s$

where d_1 is the density of the solute; carbon tetrachloride

d_2 is the density of the mobile phase, heptane

and X_s is the concentration of the solute to be detected

Thus, for the example given

$$\Delta d = (1.595 - 0.684) \times 10^{-6}$$
$$= 9.11 \times 10^{-7}$$

Now the coefficient of cubical expansion of heptane is approximately 1.6×10^{-3} per °C. It is therefore possible to calculate the change in temperature $\Delta \theta$ that would produce a change in density of the mobile phase that would be equivalent to the presence of carbon tetrachloride at a concentration of 10^{-6} g/ml.

$$\text{thus} \quad \Delta \theta = \frac{9.11 \times 10^{-7}}{1.6 \times 10^{-3}} \, °C$$

$$= 5.7 \times 10^{-4} \, °C$$

Now if it is assumed that a concentration of one part per million of carbon tetrachloride is just detectable, i.e., the thermal noise must be half the detectable signal, then the thermal fluctuations must be maintained at a level below 2.8×10^{-4} °C. Such temperature stability can be extremely difficult to maintain and thus limits the sensitivity that can be obtained from such a detector. Even the heat of adsorption and desorption of the solute on silica gel in the column could result in temperature changes of the order of 2.8×10^{-4} °C. The above argument applies equally to other bulk property detectors which monitor refractive index or dielectric constant and it must therefore be concluded that all bulk property detectors will have a limited potential sensitivity.

As a result of the limited sensitivity of bulk property detectors, they have also a very limited linear dynamic range, usually less than two orders of magnitude. Their potential dynamic range however is relatively large, perhaps four or five orders of magnitude, but this extends into the very low sensitivity areas of normal detector performance. Unfortunately the detector sensitivity settings on commercially available bulk property detectors do not permit the device to be used at very low sensitivities. This is a disadvantage as bulk property detectors are often very appropriate for preparative liquid chromatography where solute concentrations are high and out of the range of most of the detectors presently available. A low sensitivity, bulk property detector would be a very useful addition to the present range of commercially available detectors.

General Performance Characteristics and Areas of Application

Due to the general susceptibility of bulk property detectors to changes in ambient conditions, column flow rate and mobile phase composition (other than the presence of an eluted solute), they cannot be used with gradient elution, temperature programming or flow programming techniques. It follows that bulk property detectors are almost solely used under isothermal and isocratic conditions of development. The major areas of application are for detecting solutes that do not absorb in the UV wavelength range and where low sensitivities can be tolerated or are required. Bulk property detectors are frequently employed in gel permeation chromatography where in many cases the solutes do not contain UV chromophores and in preparative chromatography. Detectors in this class are employed in less than 20% of the total field of liquid chromatography. Most manufacturers of chromatographic equipment supply at least one type of bulk property detector usually the refractive index detector. They are available separately in modular form or incorporated as an integral part of the complete liquid chromatograph.

The detector alone is normally supplied in two modules; the first; the sensing unit contains the sensing cell, reference cell together with any necessary optical system, light source, power supplies and may contain coarse balancing controls as well. The second module usually called the control unit contains the amplifier together with the balancing and sensitivity controls. The cost of a bulk property detector varies somewhat between one manufacturer and another but although less versatile than solute property detectors there is little cost difference between the two types.

When entering the field of liquid chromatography the scientist is always faced with the problem of detector selection. The subject of detector choice will be dealt with in a later chapter but at this point it should again be emphasized that there is no ideal liquid chromatography detector; there is no detector with high sensitivity, wide linear dynamic range, good linearity, a predictable response and capable of detecting all solutes that is comparable to the flame ionization detector in gas chromatography. The liquid chromatographer needs to have at least two if not more different types of detectors available or the full versatility of the technique will not be realized. It is therefore recommended that one of the detectors available should be a bulk property detector, preferably the refractive index detector and particularly if preparative or semi preparative chromatography is likely to be required.

Chapter 2

The Refractive Index Detector

The refractive index detector was one of the first on-line detectors to be developed and was described by Tiselius (1) in 1942. It was also the first detector to be made commercially and was at one time the only on-line detector available for general use. Since the original model of Tiselius there have been many papers published that describe different methods of refractive index monitoring. There are four general methods for measuring refractive index, the angle of deviation method, the critical angle method, the reflection method and the Christiansen method. The theory behind each of these methods will now be discussed.

Theory

The Angle of Deviation Method

When a monochromatic ray of light passes from one isotropic medium, A, into another, B, it changes its wave velocity and direction. The change in direction is called refraction and the relationship between the angle of incidence and the angle of refraction is expressed in Snell's law of refraction

$$n_B' = \frac{n_B}{n_A} = \frac{\sin P}{\sin Q}$$

where P and Q are the angle of incident light in medium A, and the angle of refraction in medium B measured from normal, n_A and n_B are the refractive indices of medium A and medium B respectively and n_B' the refractive index of medium B relative to medium A. Refractive index is a dimensionless constant that normally decreases with increasing temperature, values given in the literature are usually measured at ambient temperature using the mean value for the two sodium lines.

If a cell is constructed in the form of a hollow prism through which the mobile phase flows, a ray of light passing through the prism will be deviated from its original path and this can be focused onto a photocell. As the refractive index of a mobile phase passing through the detector changes due to the presence of solute, the angle of deviation of the transmitted light will also change; thus the amount of light falling on the photocell will be changed and modify its output. The angle of deviation method for refractive index monitoring has been commonly employed in commercial refractometers.

The Critical Angle Method

When a ray of light passes from a medium of low refractive index to one of high refractive index the refracted ray is bent away from the normal in the second medium. If the angle of the incident ray is increased the angle of the refracted ray increases until the refracted ray is parallel to the boundary surface between the two media. At this point the incident ray is totally reflected and the angle of incidence is called the critical angle. Thus if the second medium is made the mobile phase, the angle of incidence is arranged to be close to the critical angle and the reflected ray focused onto a photocell any change in the refractive index of the mobile phase will change the amount of light reflected onto the photocell. Due to the angle of incident light being close to the critical angle slight changes in refractive index will result in significant changes in the intensity of the reflected light and thus in the output of the photocell.

The Fresnel Method

This method utilizes the relationship between reflectance from an interface between two transparent media and their refractive indices as given by the Fresnel equation:

$$R = \frac{1}{2}\left[\frac{\sin^2(i-r)}{\sin^2(i+r)} + \frac{\tan^2(i-r)}{\tan^2(i+r)}\right]$$

where R is the ratio of reflected light intensity to incident light intensity

i is the angle of incidence

and r is the angle of refraction

The Fresnel Method

Now $\dfrac{\sin i}{\sin r} = \dfrac{n_1}{n_2}$

where n_1 is the refractive index of medium 1
and n_2 is the refractive index of medium 2

Thus, if medium 2 represents the liquid eluted from the column then any change in n_2 will result in a change in R and thus a measurement of R could determine changes in the value of n_2 due to solute being present in the eluent.

Conlon (2) utilized the principle to develop a practical refractive index detector. His device is now obsolete and cannot be used with present day high efficiency columns, however, it will be described as it illustrates the principle of the Fresnel method and also demonstrates the general

Figure 1

Detector Cell Utilizing the Fresnel Principle

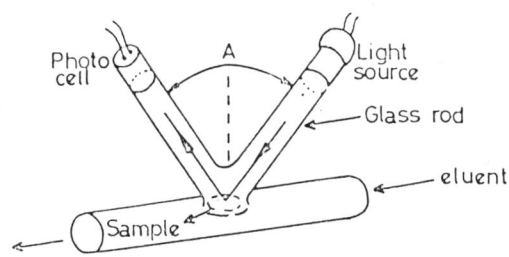

arrangement of a refractive index detector. A diagram of his detector is shown in figure 1. The sensing element consists of a rod prism sealed into a tube through which the column eluent flows. The rod prism was made from a 6.8 mm glass rod, 8 to 10 cm long, bent to the correct optical angle and an optically clear flat ground on the apex of the bend as shown in figure 1. The optical flat was then sealed into a window of a suitable tube which acted as a flow-through cell. The photocell is arranged to be one arm of a Wheatstone Bridge and the out of balance signal fed to a suitable recorder. Another photocell can be placed so as to receive light directly from the light source and if this is situated in the reference arm of the Wheatstone bridge it can help to compensate for slight variations in the light source due to voltage fluctuations.

The disadvantages of this system are that it does not contain a reference cell to compensate for temperature changes or changes in solvent composition, and the relatively large cell volume would not permit the high efficiencies obtained from present day columns to be realized.

The Christiansen Effect

This method of refractive index monitoring arose from the work of Christiansen on crystal filters (3,4). If a cell is packed with particulate material having the same refractive index as the mobile phase passing through it light will pass through the cell with little or no refraction or scattering. If, however, the refractive index of the mobile phase changes there will be a refractive index difference between the mobile phase and that of the granular packing. This difference in the refractive index results in light being refracted away from the incident beam and thus the intensity of the transmitted beam is reduced. If the transmitted beam is focussed onto a photocell and the refractive indices of the mobile phase and packing initially matched, then changes in refractive index resulting from a solute in the mobile phase will cause scattering and reduction in the output from the photocell. In practice as the optical dispersion of the two media is likely to differ, the refractive indices will only match at one particular wavelength and thus the fully transmitted light will be largely monochromatic. It follows that a change in refractive index of the mobile phase will change both the intensity and the wavelength of the transmitted light falling onto the photocell.

This device is manufactured as a commercial detector by the Gow-Mac Company, who claim it has a sensitivity of 10^{-6} R.I.U. at a signal to noise ratio of two. This is equivalent to a sensitivity of about 9×10^{-6} g/ml benzene (RI 1.501) in heptane (RI 1.388). The cell volume was kept to 8 μl and as the cell was packed with particulate material the band dispersion in the cell was minimal. Cells packed with appropriate materials to cover the refractive index range of 1.31 to 1.60 can be supplied. This detector is relatively inexpensive but suffers from the disadvantage that cells with packings of the appropriate refractive index must be used for each type of mobile phase employed. Close matching of the refractive indices of the cell packing and the mobile phase can be achieved by the use of mixed solvents and by adjustment of the cell temperature.

A Commercial Example of the Refractive Index Detector

The modern refractive index detector is the result of considerable research and development carried out over the many years perhaps starting with the work of Zaukelies and Frost (5) and Vandenheuvel and Sigus (6) and

Figure 2
The Waters Associates Differential Refractometer

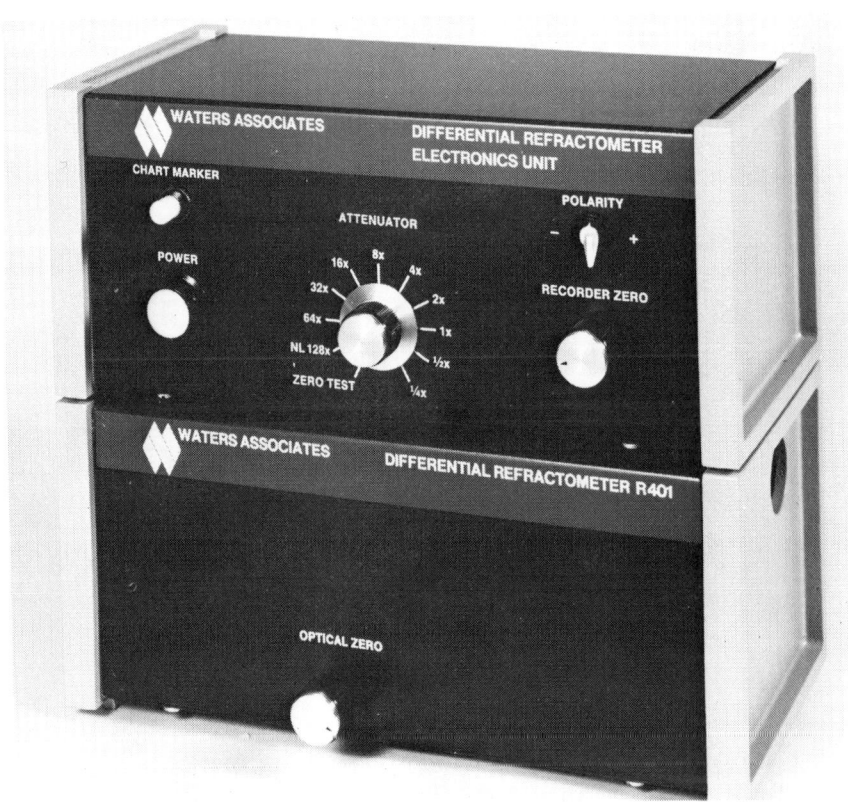

presently carried on by the effort of scientists in the research and development laboratories of the major instrument companies. An example of a modern refractive index detector is the R-400 manufactured by Waters Associates. A photograph of the instrument is shown in figure 2 and a schematic diagram that illustrates the principle of operation is shown in figure 3.

The R-400 differential refractometer measures the deflection of a light beam due to the difference in refractive index between the sample and reference liquids in a single compact sample cell. A beam of light from the incandescent lamp passes through the optical mask, which confines the beam within the region of the cell. The lens collimates the light beam and the parallel beam passes through the cells (containing the sample and reference liquids) to the mirror. The mirror reflects the beam back through the sample and reference cell to the lens, which focuses it onto a photocell.

The location of the focused beam rather than its intensity is determined by the angle of deflection resulting from the difference in RI between the

Figure 3

A Schematic Diagram of the Waters Refractometer Detector

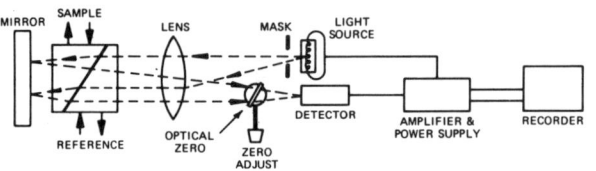

two parts of the cell. As the beam changes location on the photocell, an output signal is generated. This signal is amplified and provides an output to a meter or recorder. The optical zero glass deflects the beam from side to side to adjust for zero output signal.

By using the deflection principle of refractometry , the R-400 is able to use one sample cell throughout the entire refractive index range from 1.00 to 1.75.

An optical block and heat exchanger are provided to bring the liquid temperature to the temperature of the cell at all flow rates normally encountered. An example of a chromatogram obtained from the detector is shown in figure 4.

The specifications of the detector are given in Table I. Data for the Response Index, Linear Dynamic Range, Detector Response, unfiltered noise level or detector sensitivity were not provided in the manufacturers literature. The values for the detector response, noise level and detector

Figure 4

Chromatogram from the Waters Refractometer Detector

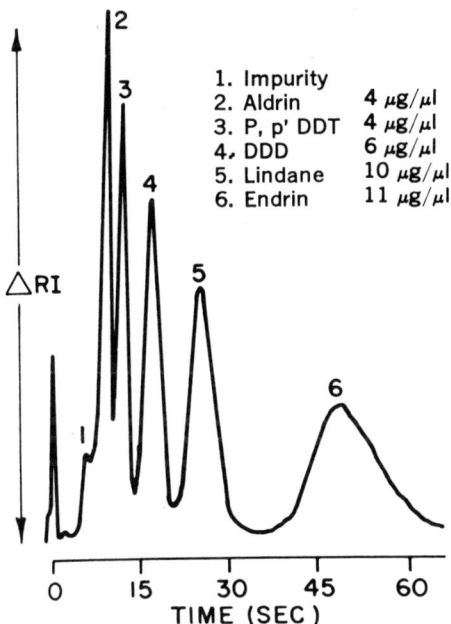

sensitivity were determined on a purchased instrument and included in the table. The detector is also available fitted with a modified cell for use in preparative liquid chromatography having a volume of 70 µl. The detector has been used in the author's laboratory on a preparative liquid chromatograph for more than two years at which time it has consistently maintained its specifications.

Table I

Specification of the Waters Associates R 401 Refractive Index Detector

Dynamic Range (D_R)	6×10^{-6} - 3000×10^{-6} RI units F.S.D. at attenuation ranges of 1/4 - 128 5.3×10^{-5} g/ml - 2.7×10^{-2} g/ml of benzene in heptane
Response Index (r)	Not given
Linear Dynamic Range (D_L)	Not given
Detector Response (R_C) *	1.85×10^5 mv/g/ml for benzene in heptane
Noise Level (N_D) *	0.52 mv at attenuation of unity
Detector Sensitivity (S_C) *	5.61×10^{-6} g/ml 6.3×10^{-7} RI Units for benzene in heptane

Cell Shape

Cell Dimensions	a = 0.118 in.	b' = 0.0625 in.
	b = 0.100 in.	a' = 0.032 in.
	c = 0.0625 in.	
	d = 0.3125 in.	
	Total V = 10 µl.	

Connecting Tube	l_T 36.5 cm
	d_T 0.23
	v_T 1.5 µl

Amplifier Time Constant (T)   Not given

The minimum value given in the dynamic range does not compare with the value for the sensitivity as at attenuations of 1/2 and 1/4, a filter circuit is introduced which reduces the noise. The time constant of the filter is not given and therefore its effect on peak dispersion cannot be assessed.

*   Determined.

Applications of Refractometer Detector

The refractive index detector will detect all substances that have a significantly different refractive index from that of the mobile phase. It is, however, one of the least sensitive detectors. As it is a bulk property detector, it is sensitive to changes in composition of the mobile phase and therefore, cannot be used with gradient elution. Further, as the refractive index of liquid varies with both temperature and pressure it cannot be used with temperature programming or flow programming techniques. It follows that the refractive index detector is almost exclusively used with isocratic conditions of development. As a result of its relatively low sensitivity it is frequently employed as the detector in preparative liquid chromatography. The detector is fairly simple to use and is sometimes employed in the preliminary scanning of samples. It is frequently used as the detector in gel permeation or exclusion chromatography. The refractive index detector is particularly valuable in the size separation of polymers where, providing the polymer has more than ten monomer units, the refractive index is directly proportional to the concentration of polymer and practically independent of the molecular weight. In a liquid chromatography laboratory that has to handle large numbers of different samples, it is desirable to have a refractive index detector available for special applications although it is likely to be employed for only a relatively small proportion of the total samples separated.

REFERENCES

1. A. Tiselius and D. Claesson, Arkiv. Kemi Mineral. Geol., 15B (No. 18) (1942)
2. R. D. Conlon, Rev. Sci. Instrum., 34 (1961) 1418.
3. C. Christiansen, Ann. Phys. Chem., 23 (1884) 298.
4. E. M. Chanot and C. W. Mason, Handbook of Chemical Microscopy, Vol. 1, Wiley, New York, 3rd ed., 1958, pp. 101 and 189.
5. D. Žaukelies and A. A. Frost, Anal. Chem., 21 (1949) 743.
6. F. A. Vandenheuvel and E. Sipas, Anal. Chem., 33 (1961) 286.

Chapter 3

The Dielectric Constant Detector

Theory

Under the influence of small fields, electrons move quite freely through conductors, whereas in insulators or dielectrics these fields displace the electrons only slightly from their equilibrium positions. As an electric field acting on a dielectric causes a separation of positive and negative charges, the field is said to polarize the dielectric. The polarization can occur as a result of two effects: the induction effect and the orientation effect. An electric field always induces dipoles in molecules on which it is acting, whether or not they contain dipoles to begin with. If the dielectric does contain molecules that are permanent dipoles, the field tends to align these dipoles along its own direction. As a result of the induction or orientation it is found experimentally that when a dielectric is introduced between the plates of a capacitor the capacitance is increased by a factor ε, called the dielectric constant. Thus if C_o is the capacitance with a vacuum between its plates, the capacitance with a dielectric is $C = \varepsilon C_o$. In this way the dielectric constant of a substance can be defined. Due to the electro magnetic nature of light, it is also affected by the dielectric constant of the medium it passes through. It follows that the refractive index of a substance is a similar property to dielectric constant and in some circumstances is a direct function of it. For example for non polar substances or mixtures of non polar substances the relationship between the dielectric constant ε and the refractive index of the substance or mixture, n, is given by

$$\varepsilon = n^2$$

For semi polar substances or mixtures or semi polar substances and non polar substances the above equation has to be modified to the following form

$$\frac{\varepsilon - 1}{\varepsilon + 2} = \frac{n^2 - 1}{n^2 + 2}$$

For polar substances or mixtures of polar substances and semi polar substances, however, the relationship breaks down and there is no simple function that describes refractive index in terms of dielectric constant.

In general the more polar the substance the larger is its dielectric constant. This is always true for substances having monofunctional groups and generally true for substances having more than one functional group but there are exceptions. For example, dioxane with two ether groups has a fairly low dielectric constant although it is a very polar solvent. The low value for the dielectric constant results from the fact that the two dipoles are electrically in opposition and thus partially neutralize the effect of each others charge. This effect is worth considering when choosing the mobile phase for use with the dielectric constant detector.

In chromatography the mobile phase is usually less polar than the solutes being eluted as they need to be retained on the column to effect a separation. Thus the presence of a solute in the mobile phase will, in general, change the dielectric constant of the mobile phase. If a device is situated at the end of the column that responds to changes in dielectric constant such a device can act as a chromatographic detector.

In practice the sensory element usually takes the form of a cylindrical or parallel plate condenser. To maintain column efficiency the volume of the condenser has to be very small and, as the sensitivity of the device is directly related to capacity of the condenser, the plates have to be very close together.

The capacity (C) of a parallel plate condenser is given by

$$C = \frac{0.0885 \, (N-1) A \varepsilon}{d}$$

where ε is the dielectric constant
N is the total number of plates
A is the area of the plates in system in cm^2
and d is the distance between the plates in cm

The capacity of a cylindrical condenser is given by

$$C = \frac{0.2416\varepsilon l}{\log r_1/r_2}$$

where l is the length of the cylinder in cm

r_1 is the radius of the outer cylinder in cm

r_2 is the radius of the inner cylinder in cm.

The impedance (i), of a condenser, which is in effect its resistance to an alternating electrical supply is given by

$$i = 1/2\pi fC$$

where f is the frequency of the applied AC potential.

A simple circuit for use with a dielectric constant detector is shown in figure 1. An AC potential resulting from the current passing through the detector capacitor C develops a voltage across R_1 which is rectified by the diode and the resulting DC potential across R_2 is fed to an amplifier or

Figure 1

Simple Dielectric Constant Detector Circuit

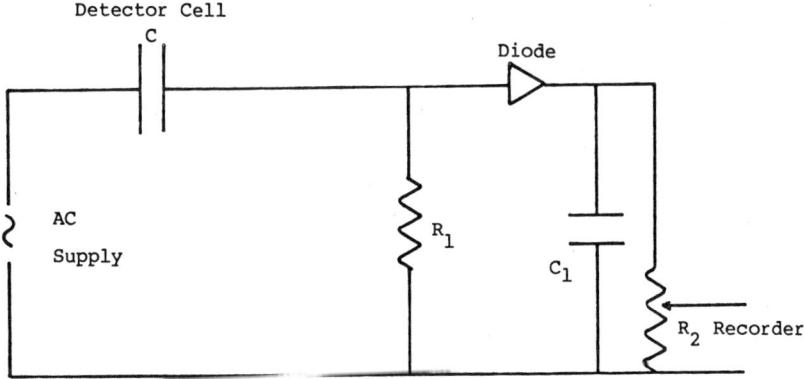

recorder. C_1 is solely a smoothing capacitor. The circuit merely illustrates the principal involved and is of little value in practice because it is not compensated for changes in the electrical supply and, furthermore, provides a standing current across R_2 irrespective of the presence of a solute. This standing current will give noise from fluctuation in supply voltage or frequency even if backed off by a suitable DC potentiometric system.

A more appropriate circuit to use is an AC bridge, the detector condenser being situated in one arm of the bridge; however the type of bridge that is appropriate depends on the design of the detector cell. If the capacity of the cell is reasonably large (100 pF or more) a Wein Bridge can be used, if small (1-10 pF) then it is more appropriate to use the Schering Bridge; both these bridge systems will be described but before this is done some discussion on the balancing of an AC bridge is appropriate. No capacitor is ideal insomuch that there is always some resistance component associated with it. If the plates of the detector cell are not insulated from the mobile phase and the mobile phase is conducting then the resistance component can obviously be very large. Now the voltage developed across the resistive component of the condenser is out of phase with the voltage developed across the capacity component. It follows that to balance the bridge the two components have to be balanced separately or no null balance point will be found. It should also be noted that if the eluted solute changes both the dielectric constant and the resistance of the mobile phase, as in the case of ionized solutes, then both the resistance and the capacity of the detecting cell will be changed and thus could provide a greater response than the effect of either alone. Thus a dielectric constant detector that responds to both dielectric constant and conductivity of the mobile phase would be very sensitive to ionic solutes and this could be achieved by using uninsulated plates in the detector capacitor. The theory of the combined effects of electrical conductivity and dielectric constant on detector output has been discussed in detail by Haderka (1) and workers interested in this aspect of dielectric constant detection are recommended to read his original paper.

A diagram of the Wein Bridge used for detectors having capacities of 100 pF or more is shown in figure 2. ABEF is the bridge with an AC potential applied across A and E. The out of balance signal is sensed across B and F by D which can be an AC amplifier and rectifier feeding a potentiometric recorder. C and r represent the capacity and resistance of the detector cell. C_o is a standard reference capacitor or can be the capacity of the reference cell if one is employed; R is a fixed resistor and R_o and r_o are variable resistances for obtaining balance. Balancing is achieved by

Figure 2

The Wein Bridge

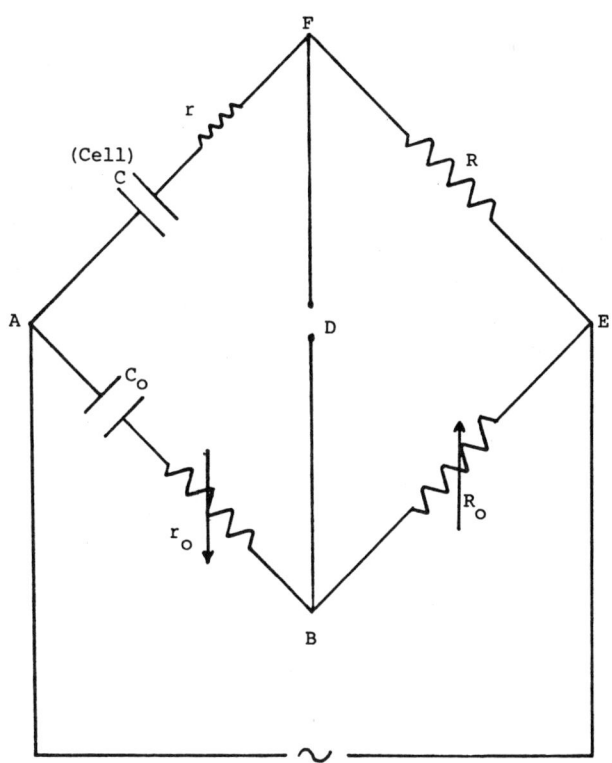

iterating adjustments of R_o and r_o for zero output across FB. The conditions of balance are.

$$\frac{R}{R_o} = \frac{r}{r_o}$$

or

$$\frac{C}{C_o} = \frac{R_o}{R}$$

74   Liquid Chromatography Detectors

Any change in C or r due to the presence of solute in the detecting cell will result in an off balance signal across FB.

For small capacity cells the Schering Bridge can be used which was originally designed for testing cables. A diagram of the Schering Bridge is

Figure 3

The Shering Bridge

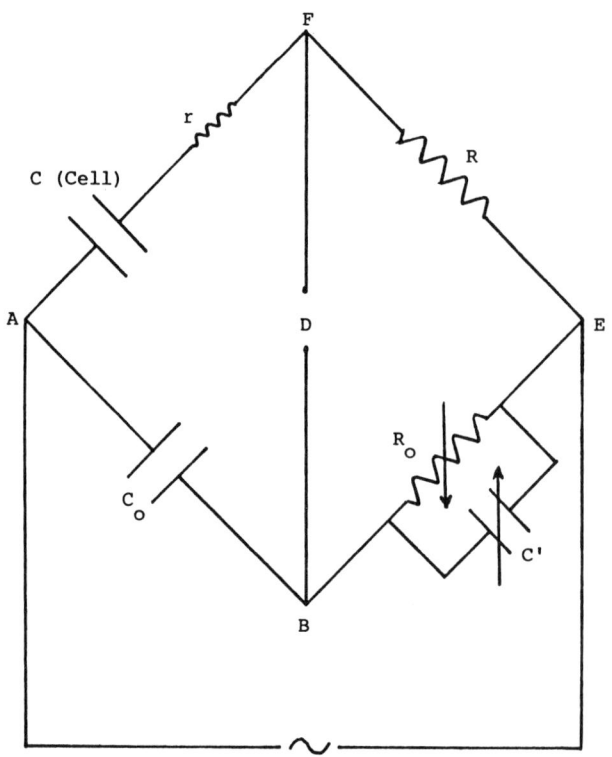

shown in figure 3. The alternating supply is applied across AE and the out of balance signal taken across FB in a similar manner to the Wein Bridge, C and r represent the capacity and resistance of the detecting cell, R and C_o are the reference resistance and capacitor and if a reference cell is employed C_o can be the capacity of the reference cell. R_o and C' are a variable resistance and capacity respectively for obtaining balance.

Balancing is achieved by the iterative adjustment of R_o and C' and, when balanced, any subsequent change in C or r resulting from solute in the detector cell will produce an off balance signal across FB. The conditions for balance are as follows

$$\frac{C}{R_o} = \frac{C_o}{R}$$

and $\quad \dfrac{r}{R} = \dfrac{C'}{C_o}$

It should be emphasized that the resistance component of the cell across the detector capacity reduces the bridge sensitivity or makes the initial balancing procedure tedious. To reduce this effect, the plates of the detecting condenser are often well insulated from the mobile phase thus eliminating the resistive component of current.

Another method of measuring changes in dielectric constant is to make the condenser containing the dielectric part of an oscillator circuit. For example, if the capacity is connected in parallel with an inductance and made the frequency source of an oscillator, the frequency will depend, among other things, on the capacity of the condenser. If the output is heterodyned, with the output of a stable reference oscillator and balanced, then the difference frequency resulting from a change in capacity due to the presence of a solute can be passed to a discriminator, rectified and the DC output fed to a recorder. This alternative method of measuring changes in dielectric constant can be extremely sensitive. However, it is not the electrical measuring system that determines the overall detector sensitivity, but the limitations inherent in bulk property detectors that have already been discussed.

One of the early dielectric constant detectors to be described was that by Grant (2), the cell consisted of a flattened glass bulb containing two thin platinum electrodes, 1 sq. cm in area and 4 mm apart. The plates were immersed in, and in contact with, the mobile phase and were fused to the wall of the bulb to prevent vibration giving rise to detector noise. The cell holder was made of 2.5 cm diameter brass tube and was connected to one plate of the condenser and to earth to provide electrical stability. The capacity measuring unit and detector cell derived its power from a stabilized power supply unit. The volume of the cell, was fairly large (2-3 ml) and this would obviously impair the resolution obtained from modern microparticulate columns of high efficiency.

Johansson and Karrman (3) described a dielectric constant system that used a different type of cell in which the plates were kept out of contact with the mobile phase. As well as improving the performance of the measuring unit, the authors claimed that the isolation of the detector plates prevents the eluted solutes from being decomposed or adsorbed on exposed metal surfaces in the cell. The cell consisted of a straight glass tube that could be connected to the column by means of a ground glass spherical joint. The plates were semi-cylindrical in shape, 3 mm radius and 70 mm long. To reduce wall effects the glass tube was chosen to have walls as thin as possible (0.3 mm thick). The glass tube and semi-cylindrical plates were enclosed in a brass tube which was connected to one of the plates and earthed to improve stability. Again, however, the cell volume

Figure 4

Schematic Section of Detector Cell

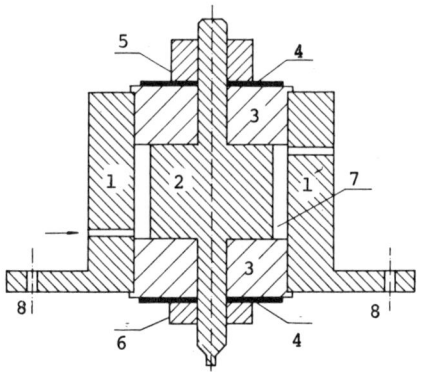

1 = Detector body serving as earthed plate of the capacitor; 2 = core serving as second capacitor plate; 3 = Teflon sealing; 4 = washer; 5, 6 = nuts; 7 = space for the dielectric; 8 = hole for tightening screw. All parts are made of brass. Eluate passage is marked with arrow.

was large, ca. 160 µl and this could not be used with present day microparticulate columns. More recently Vespalec and Hána (4,5) described a detector suitable for detecting substances having no electrical conductivity and a diagram of their cell is shown in figure 4. The cell consisted of two brass cylinders (between which the mobile phase passed) having radii of 9.85 and 10 mm respectively and were 4.96 mm long which gave a detector volume of about 12 µl. The detector had a sensitivity of about 10^{-6} g/ml for acetone (ε = 20.7) in hexane; the linear range was stated to be about 10^4 and so would be very suitable for preparative liquid chromatography.

Poppe and Kunysten (6) designed a dielectric constant detector which included a reference cell for temperature compensation. The cell consisted of two stainless steel plates 2 cm x 1 cm x 1 mm separated by a gasket - 50 micron thick, the two cells, the reference cell and the detecting cell were identical and clamped back to back thus sharing a common electrode. The stated sensitivity was 10^{-6} g/ml for chloroform (ε = 4.81) in isoctane, the cell system, however, was found to be very sensitive to pressure changes even when constant flow pumps were employed. Fluctuations in the outlet pressure of the column were thought to deform the plates and thus produce sporadic noise. Another detector described by Ooguri employed a capacitance cell, cylindrical in form with a central concentric electrode. The cell was used as the capacity of a tuned circuit that was heterodyned with 9 MHz oscillator, the difference frequency being rectified and fed to a suitable recorder. The volume of the cell was stated to be 10 µl and 10 µg of material was detectable. However, the sensitivity in terms of concentration was not given.

Generally the same comments apply to dielectric constant detectors as to refractive index detectors. They have relatively low sensitivity and are best employed where chromatographic conditions are kept constant. They are not suitable for use with gradient elution, or where temperature programming or flow programming is used. If carefully calibrated, they can be used for quantitative analysis but they do not have a linear response over a significant concentration range. The main areas of application for dielectric constant detectors are the same as for refractive index detectors and if a suitable commercial dielectric content detector was available it could be used as an alternative to the refractive index detector. At present, to the best of the author's knowledge, there is no dielectric detector commercially available. The dielectric constant detector is one of the detectors that could be designed to have a very small dead volume if, in the future, such a detector was required for use with microbore columns.

REFERENCES

1. S. Haderka, J. Chromatogr., 91 (1974) 167.
2. R. A. Grant, J. Appl. Chem., 8 (1959) 136.
3. G. Johansson and K. J. Karrman, Anal. Chem., 8 (1958) 1397.
4. R. Vespalec and K. Hána, J. Chromatogr., 65 (1972) 53.
5. R. Vespalec, J. Chromatogr., 108 (1975) 243.
6. M. Poppe and J. Kunysten, J. Chromatogr. Sci., 10 (1972) 16A.
7. F. W. Karasek, Res./Develop., 26 (1975) 34.

Chapter 4

The Electrical Conductivity Detector

An acid, base or salt, when dissolved in water ionizes into charged ions. If the acid, base and salt are represented by HA, BOH and BA respectively the ionization can be described by the following equations

$$HA \rightarrow H^+ + A^-$$
$$BOH \rightarrow B^+ + OH^-$$
$$\text{and } BA \rightarrow B^+ + A^-$$

If the acid or base is weak, ionization is incomplete and an equilibrium condition occurs where

$$k_A(HA) \rightleftharpoons (H^+) + (A^-)$$

$$k_B(BOH) \rightleftharpoons (B^+) + (OH^-)$$

and k_A and k_B are the dissociation constants of the acid and base respectively.

In any event some ions are produced, their concentrations being dependent on both the original concentration of the acid or base, their respective dissociation constants and the physical properties of the solvent. Under the influence of a potential gradient the ion in solution can carry an electric charge and thus if a voltage is applied across two electrodes situated in the solution, a current will flow between the electrodes and the solution is said to be conducting. It is fairly obvious that such a system could be used as a method of detecting ionic species in liquid chromatography but before discussing the design of electrical conductivity detectors it would be advantageous to consider in more detail the properties of ions in solution.

Early investigations into the electrical properties of solutions containing ions were somewhat frustrating as DC voltages were employed which resulted in the production of hydrogen and oxygen at the electrodes. This effect, called polarization, changed the electrical resistance of the system at the electrodes and produced confused results. It was not until AC potentials were employed at the electrodes which rendered polarization effects insignificant that the true electrical properties of solutions could be identified. It follows that if the measurement of the electrical properties of ionic solutions is to be used effectively as a liquid chromagraphy detecting system, AC potentials must be applied to any electrode configuration employed.

Experiments using AC potentials across the electrode system demonstrated that ionic solutions obeyed Ohms Law i.e.

$$V = RI$$

where V is the applied potential in volts
I is the current flowing between the electrode in amps
or R is the electrical resistance between the electrodes in ohms

Generally in physical chemistry the conductivity of the solution is of greater interest and is equivalent to the reciprocal of the specific resistance of a solution. The specific resistance of a solution is numerically equivalent to potential in volts across the faces of a centimeter cube of the solution when carrying a current of one amp. In electrical conductivity detectors it is the resistance of the solution that is actually monitored and it is the change in electrical resistance of the mobile phase in the presence of a solute that provides the output from the detector. For this reason, despite the term electrical conductivity detector the functioning of the detector will be considered in term of resistance measurement.

The chromatographer is so familiar with water as the medium in which the ionization of solutes takes place that he tends to overlook the possibility of other solvents. Liquid ammonia, liquid sulphur dioxide, hydrogen fluoride, hydrogen sulphide and hydrogen cyanide are some typical examples of alternative ionizing media. Compared with water these substances ionize in the following manner

Substance	Cation		Anion
$2H_2O$	$(H_3O)^+$	+	OH^-
$2NH_3$	$(NH_4)^+$	+	NH_2^-
$2SO_2$	$(SO)^{2+}$	+	SO_3^{2-}
$2HF$	$(H_2F)^+$	+	F^-
$2H_2S$	$(H_3S)^+$	+	SH^-
$2HCN$	$(H_2CN)^+$	+	CN^-

Liquid ammonia and liquid sulphur dioxide appear attractive possibilities for chromatographic purposes as both these substances, particularly liquid sulphur dioxide are very good solvents for organic substances. Furthermore they are likely to ionize specific organic materials that otherwise do not normally ionize in water or are insoluble in water. Those substances hitherto not ammenable to separation by ion exchange chromatography might be efficiently separated by ion exchange chromatography using liquid sulphur dioxide as the mobile phase. Using liquid sulphur dioxide as an ionizing mobile phase also opens up possibilities for the design of entirely new forms of ion exchange media.

The liquid chromatographer, at the first thought of such a system, may well be dismayed at the problems of toxicity and mechanical handling that appears to accompany the use of such materials. However in the early days of refrigeration entirely sealed systems were manufactured successfully for use in domestic environments employing both ammonia and liquid sulphur dioxide as the refrigerating liquid. It follows that the development of sealed liquid chromatographic systems incorporating such materials for use by scientists in a laboratory environment appears distinctly possible. To quote Professor C. Giddings from a lecture on chromatography delivered in the early 60's; "If there are advantages to be gained, man has usually shown himself capable of surmounting any practical problems that prevent him from realizing those advantages".

The first effective conductivity detector to be described was that of Martin and Randall (1). Improved cell designs have been described by Harlan (?), Sjoberg (3) and more stable and sensitive electronic circuits for use with conductivity detectors have been discussed by Avinzonis and Fritz (4) and Berger (5). Scott, Blackburn and Williams (6) inserted electrodes in the walls of a column to monitor changes in band dispersion along a chromatographic column by conductivity measurement.

An electrical conductivity detector cell consists of a small chamber containing two electrodes across which an AC potential is applied. The volume of the cell should be as small as possible to minimize dispersion. The electrodes should be constructed of an inert conducting material such as stainless steel, gold or platinum and they should be well insulated from one another so that only the resistance of the liquid between them is measured. The electrical capacity of the cell should also be kept to a minimum by ensuring that the surface area of the electrodes is as small as possible. If the cell has significant electrical capacity and a bridge circuit is used, then as an AC potential has to be employed to eliminate polarization, electrical balance will be difficult to achieve due to the out of phase

Figure 1

Wheatstone Bridge for Use with the Conductivity Cell

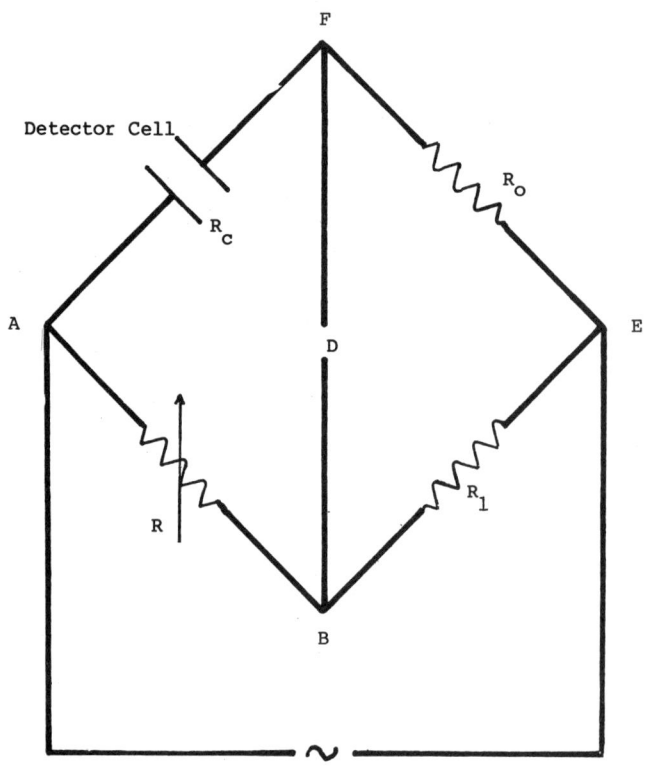

capacity current across the bridge. The resistance of an ionic solution can change significantly with temperature and it may be necessary to thermostat the cells.

The measuring circuit is usually a simple Wheatstone Bridge, a diagram of which is shown in figure 1. A stable AC voltage having a frequency of about 1 kHz is applied across AE and the output across FB taken, usually via an AC amplifier, fed to a bridge rectifier and thence to the potentiometric recorder. An attenuating network is often situated subsequent to the bridge rectifier. The bridge is initially balanced by adjustment of R and under conditions of balance

$$\frac{R_c}{R} = \frac{R_o}{R_1}$$

where R_c is the resistance of the detector cell and R_o is chosen to approximately match R_c when no solute is present in the cell. A specific value for R_o cannot be given as the value of R_o will depend on the specific resistance of the mobile phase which will vary considerably depending on the concentration of any buffer employed for each particular chromatographic development. If the detector cell has significant capacitance a balance will not be attenable by adjustment of R due to the out of phase capacity current. Under such circumstances a Wein or Shering bridge should be used as discussed in the previous chapter.

A Commercial Example of an Electrical Conductivity Detector

The Conductomonitor manufactured by Laboratory Data Control is a typical example of a commercially available electrical conductivity detector. A photograph of the instrument is shown in figure 2. It consists of two modules, the conductivity cell assembly and the control unit. The control unit has five range settings for conductivity or resistance (10, 100, 1000, 10,000, 100,000 pc mho/cm or 100,000, 10,000, 1000, 100 ohms/cm) and thus can be used with mobile phases having a wide range of conductivity or specific resistance. This provides adequate flexibility in the choice of buffer concentration in ion exchange chromatography. Mobile phases containing significant concentrations of buffer salts will have a relatively low specific resistance and thus the presence of solute ions only slightly modify the overall inter electrode resistance. For this reason at

Figure 2

The Laboratory Data Control ConductoMonitor

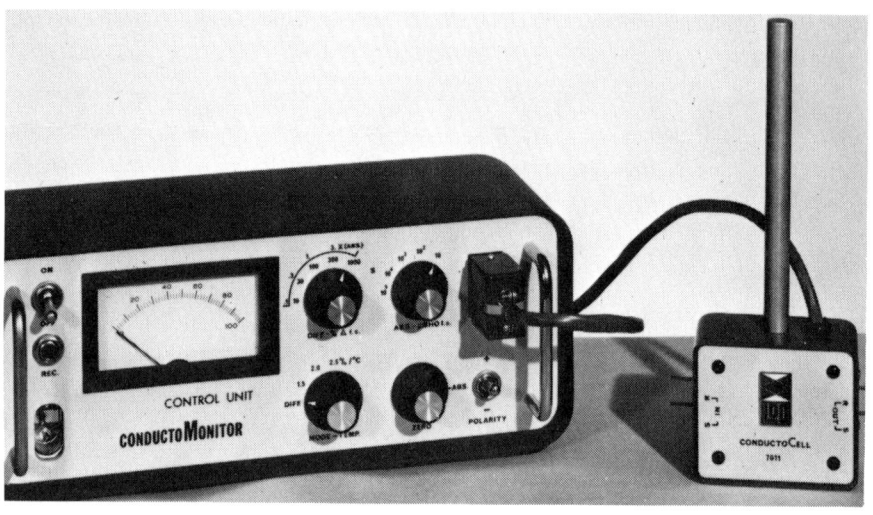

high buffer concentration, low values of attenuation are usually employed. The control unit provides for levels of attenuation of 0.1, 0.3, 1.0 and 3.0. Zero suppression equivalent to ten times the respective range setting is also provided so that high out of balance standing currents can be tolerated.

A diagram of the cell assembly is shown in figure 3. The unit contains reference and sample cells which help to compensate for changes in ambient temperature and also permits the differential mode of operation. In the differential mode of operation the resistance of the sample cell is compared with that of the reference cell and as the two cells share a common electrode the two cells become the two arms of the Wheatstone bridge. The differential mode provides much higher sensitivity particularly when high buffer concentrations are used in the mobile phase and the specific resistance is low. In the absolute mode the resistance of the sample cell is compared to a fixed resistance which replaces the reference cell in the other arm of the bridge. The electrodes in each cell consist of two stainless steel tubes insulated from one another by polytetrafluoroethylene.

Figure 3

The Conductomonitor Cell Assembly

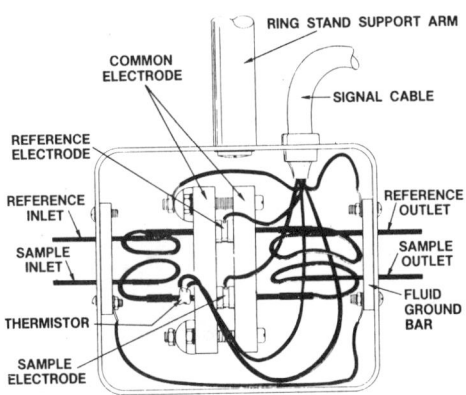

The actual cells themselves have volumes of only 2.5 µl; however, the tubes connecting them to the column are also made of polytetrafluoroethylene but are several centimeters long having a volume of 25 µl giving a total detector volume of 27.5 µl. The long length of tubing is apparently necessary to provide as high an impedance as possible between the detector cell and the column which will be at electrical earth. As the mobile phase will be conducting if this impedance is not made sufficiently high, the 'earth loop' to the detector amplifier and recorder will produce serious instability. However the extraneous connecting tube volume could be significantly reduced if tubes of a reduced internal diameter were used.

The voltage supply to the cells has a maximum of 1 volt at a frequency of 2 kHz and a phase sensitive detection circuit is employed. The specifications of the detector are given in table I.

A small volume dielectric constant detector cell described by Pecsok and Saunders (7) is shown in figure 4. The column is connected with a Swagelok fitting to the detector before being packed. The short tubing connector is epoxied into a Lucite block which is epoxied and screwed to a rimmed stainless steel plate. The bottom half is identical. These two stainless plates each contact one of the two small (9/16-inch diam) disks of 80-mesh

Table I

Specifications of the Laboratory Data Control Electrical Conductivity Detector

Dynamic Range	1–100,000 µl mho/cm specific concentration range not given
Response Index	not given
Linear Dynamic Range	± 0.2% from 0.01 to 100,000 µl mho/cm specific concentration range not given
Detector Response	not given
Noise Level	0.2% of full scale deflection on all ranges
Detector Sensitivity	not given
Cell Shape	cylindrical
Cell Dimension	not given volume 2.5 µl
Connecting Tube	25 µl
Amplifier Time Constant	not given

Other Data:

Cell Constant	20 cm^{-1} nominal
Temperature Limit	30°C
Pressure Level	20 psi

(0.003-inch diam wire) platinum-rhodium wire cloth. The two platinum disks sandwich two pieces of 400-mesh nylon net. The hole running through the length of the detector is 5/32-inch diam, the same as the column tubing i.d. The neoprene gasket is 0.148 inch thick, while the combined thickness of the platinum disks, nylon disks, and stainless steel rims is 0.116 inch, leaving 0.032 inch for sealing by compression of the gasket. Electrical connection is made by coaxial cable to the stainless plates which are insulated from one another by bolting through lucite standoffs. The column is then packed

Figure 4

Low Volume Dielectric Constant Detector Cell

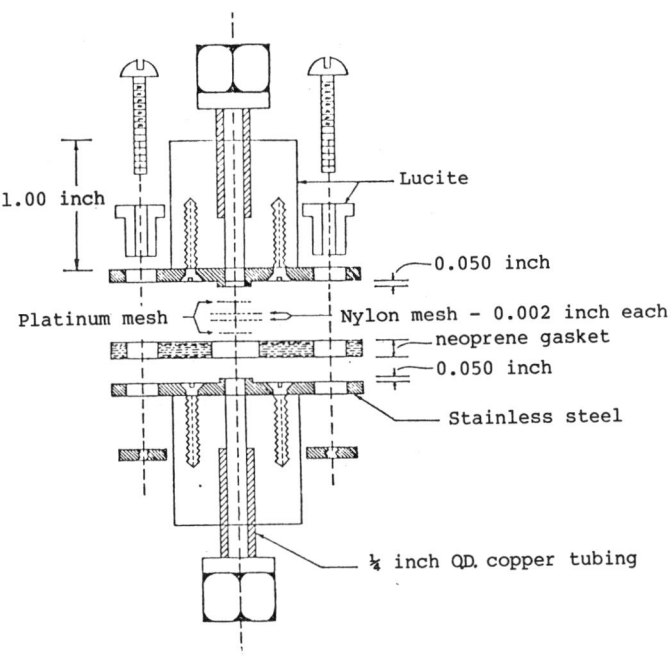

so that the first layer of packing material lies directly on the platinum mesh. The specific sensitivity of the detector was not given and the authors stated that with their particular electronics the response of the detector was non linear.

Recently Scott and Reese (8) developed a conductivity detector having a total volume of 0.6 μl. Their cell design is shown in figure 5. The electrodes consisted of two lengths of 1/16 in O.D., 0.005 in I.D. stainless steel tubing 2 mm long separated by a ring of polytetrafluoroethyl 0.002 inch thick housed in a 1/6 in nylon Swagelok union. Connections to the electrodes were made by screws threaded through the unions and tightened against the electrodes. The column connecting tube and "exit" tube were also made of 1/16 in in O.D., 0.005 in I.D. stainless steel tube which were, in turn, insulated from the electrodes using polytetrafluoroethylene rings.

The cell was used in conjunction with a Wein Bridge and the frequency of

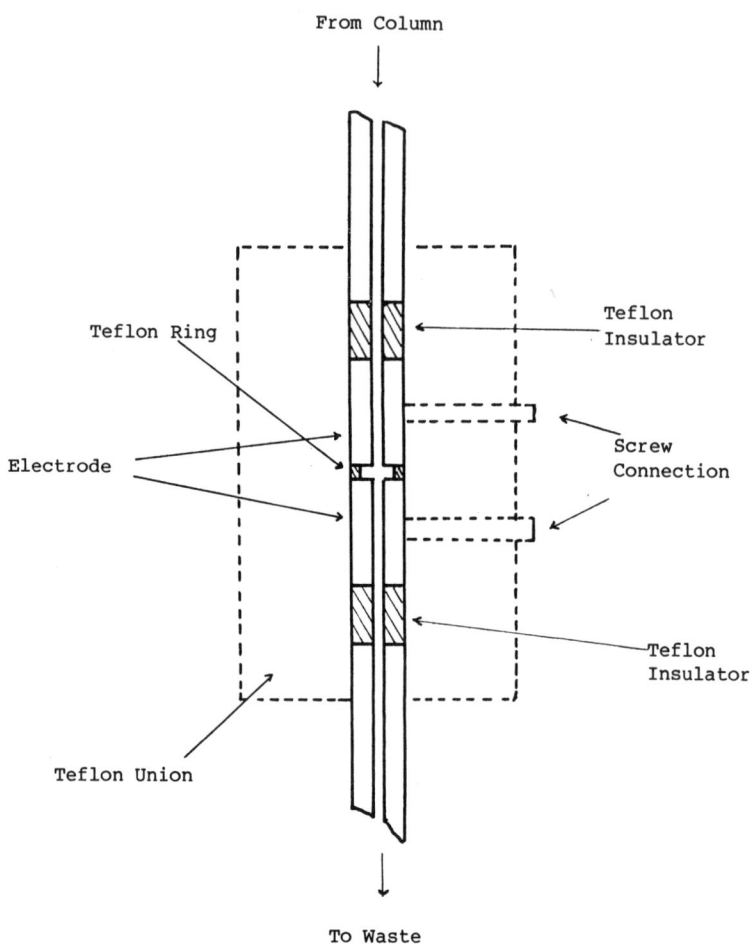

Figure 5

Sample Conductivity Cell of Small Dead Volume

the applied voltage was 1 kHz. Details for the performance of the detector will be available in the future.

The conductivity detector has a wide field of application to the detection of all solutes that produce ions in solution and thus can be used in the chromatographic separation of both organic and inorganic acids, bases and salts. When used with buffer solutions of relatively high conductivity the sensitivities realized will be of the order of 10^{-6} g/ml, characteristic of bulk property detectors. Even when the differential mode of detection is employed sensitivities much above this value are seldom realized with high conductivity mobile phases. However, when used with mobile phases of relatively low conductivity the performance of the detector tends towards that of a solute property detector and sensitivities of 10^{-8} to 10^{-9} g/ml can be obtained. The chromatogram of a mixture of acid peptides separated by exclusion chromatography on silica gel employing

Figure 6

Chromatogram of Acid Peptides by Exclusion Chromatography using a Conductivity Detector

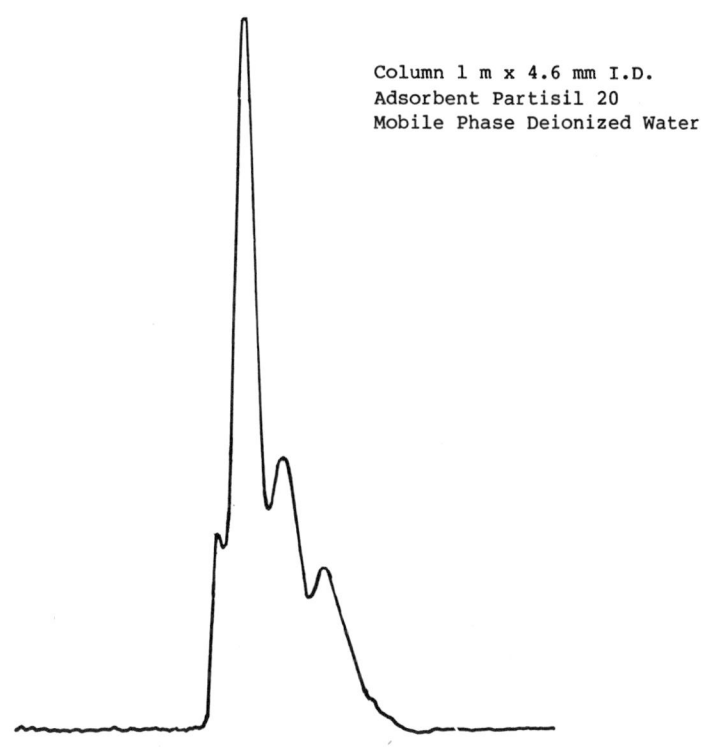

Column 1 m x 4.6 mm I.D.
Adsorbent Partisil 20
Mobile Phase Deionized Water

deionized water as the mobile phase in conjunction with the conductivity detector is shown in figure 6. In the example the mobile phase has an extremely low conductivity and the detector is in effect a solute property detector exhibiting a sensitivity of about 10^{-8} g/ml. The conductivity detector is probably the detector of choice for all ionizible solutes but to date it has been a somewhat neglected detecting system with only a few examples of its use in the literature. This may be due to the limitations in the presently available designs, the need for carefully constructed bridge circuits and amplifiers to provide a linear response or possibly workers in the field being unaware of its potentialities; nevertheless, it is likely that in the future the conductivity detector will find increasing use in the chromatography of mixtures of ionic substances.

REFERENCES

1. A. J. P. Martin and S. S. Randall, Biochem. J., 49 (1951) 293.
2. H. D. Harlan, Anal. Chem., 54 (1965) 89.
3. C. I. Sjoberg, Acta. Chem. Scand., 8 (1954) 1161.
4. P. W. Avinzonis and F. Fritz, Anal. Chem., 34 (1962) 58.
5. D. Berger, Arkiv. Kemi, 4 (1952) 401.
6. R. P. W. Scott, D. W. J. Blackburn and T. Wilkins, J. Gas Chromatogr., (1967) 183.
7. R. L. Pecsok and D. L. Saunders, Anal. Chem., 40 (1968) 1756.
8. R. P. W. Scott and C. E. Reese, in preparation.

Chapter 5

Additional Bulk Property Detecting Systems

The bulk property detectors previously described employ well established detecting methods that have been investigated by numerous scientists and are, for the most part, extensively used in the field of chromatography. There are, however, a number of bulk property detecting systems that have been described in the literature and that have either not been developed to their full potential or have a limited performance. Some of these detecting systems will be discussed in this chapter to give the reader an idea of the range of physical properties that have been examined as a possible basis for chromatographic detection. Hopefully it may also stimulate the reader to develop some of them to a greater level of sophistication and reliability.

The Density Detector

The density of the mobile phase leaving a column will change when a dissolved solute is present and thus the continuous measurement of solvent density can become an effective method of detection. Such a method has been described by Fornstedt and Porath (1) and a diagram of their detecting cell is shown in figure 1. It consists of a spherical glass plumb supported from the arm of an automatic continuously recording electrobalance (Cahn) and totally immersed in the eluent leaving the column. The plumb was 5 mm in diameter and about 0.4 g in weight. The solvent flowed round the plumb and then into the overflow tube as shown in figure 1. In order to maintain stability the chamber containing the float was situated in a thermostat and the mobile phase brought to the temperature of the cell prior to entering by passage through a metal coil situated between the column and the detector chamber.

Owing to its nature, the arrangement is inherently very sensitive to mechanical and electrical noise. Mechanical vibrations can be avoided by mounting the balance on a very heavy support. Electrical disturbances must be minimized by careful earthing and temperature gradients within or around

Figure 1

The Detector Cell of the Density Detector

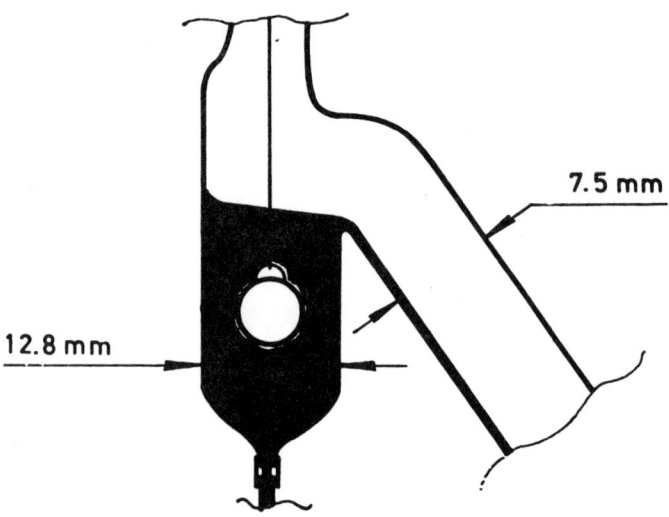

the apparatus can seriously disturb the system. The entire liquid system should be thermally insulated so that the maximum range of temperature variation is one degree above or below the prevailing ambient temperature. A further disadvantage of the system is the large dead volume of mobile phase in the detector chamber itself that is necessary to contain the glass plumb. The relatively large volume would cause very serious dispersion of the detector if employed with modern high efficiency columns. An example of a chromatogram of a mixture of t-Boc-valine and phenylalanine methyl ester separated on Sephadex LH-20 and obtained from this detector is shown in figure 2. Each peak represents 50 mg of solute and so it is seen that the detector exhibits very poor sensitivity. It will, however, detect any solute having a density different to that of the solvent but has to be used under conditions of isocratic development.

Figure 2

Chromatogram from the Density Detector

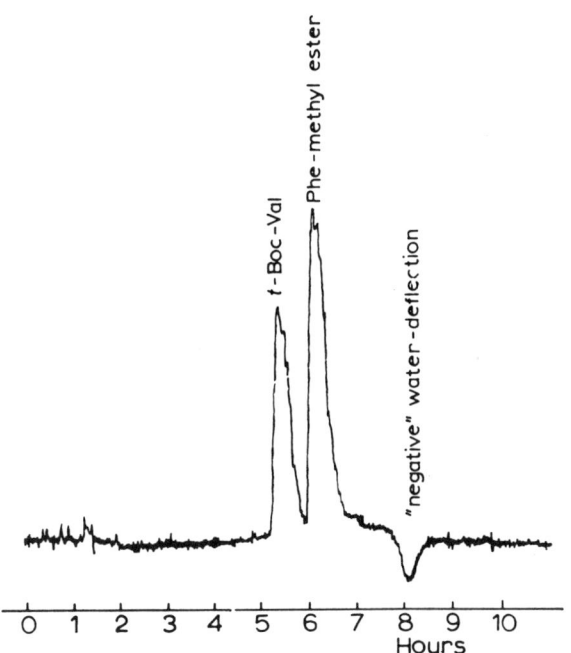

The Thermal Conductivity Detector

The presence of a solute in the mobile phase leaving a chromatographic column will change the thermal properties of the mobile phase and thus a device that continuously monitored the thermal properties of the eluent could act as an effective detector, the katharometer employed as a gas chromatographic detector functions on this principle. The thermal properties that change significantly when a solute is present in the mobile phase are thermal conductivity and specific heat. Ohzeki et al. (2) have described, what they term, a Thermal Conductivity Detector for use in liquid chromatography. It appears, however, on reading their paper that the detector is probably responding more to changes in the specific heat of the mobile phase than to thermal conductivity. A diagram of their detector

Figure 3

The Thermal Conductivity Detector

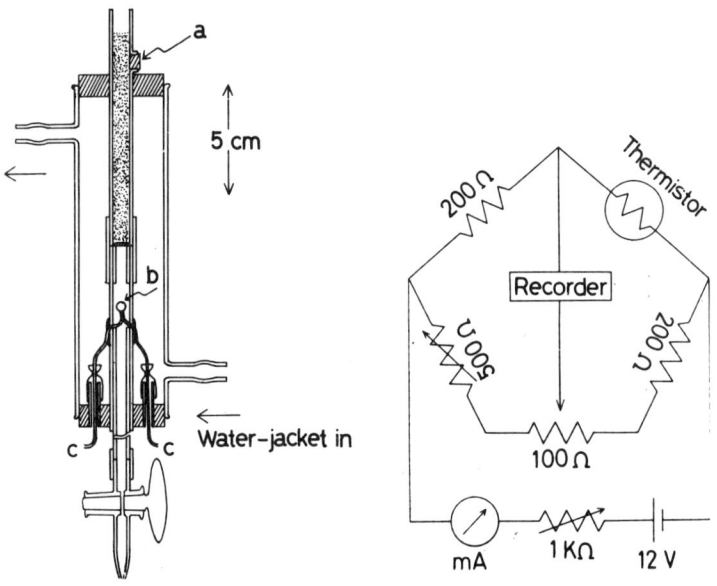

together with the bridge circuit employed with it is shown in figure 3. A thermistor is situated in the mobile phase leaving the column and is heated by a current passing through it. The thermistor comes to an equilibrium temperature and thus a constant resistance when the heat generated by the current passing through is equal to the heat lost to the mobile phase. In the presence of a solute the heat lost to the mobile phase will differ resulting in a change in temperatue of the thermistor. The resistance of the thermistor will thus also change which will unbalance the bridge and provide an output to the recorder. An example of some elution curves obtained from the detector for Blue Dextran separated on a Sephadex G-25 column are shown in figure 4. The peaks in figure 4 indicate that a very useful sensitivity can be obtained from this detector and further, the dead volume of this device could be made relatively small and thus it could be used with high efficiency columns. It would, however, have to be employed under isocratic conditions of development and as the detecting system

Figure 4

Elution Curves for Blue Dextran from the Thermal Conductivity Detector

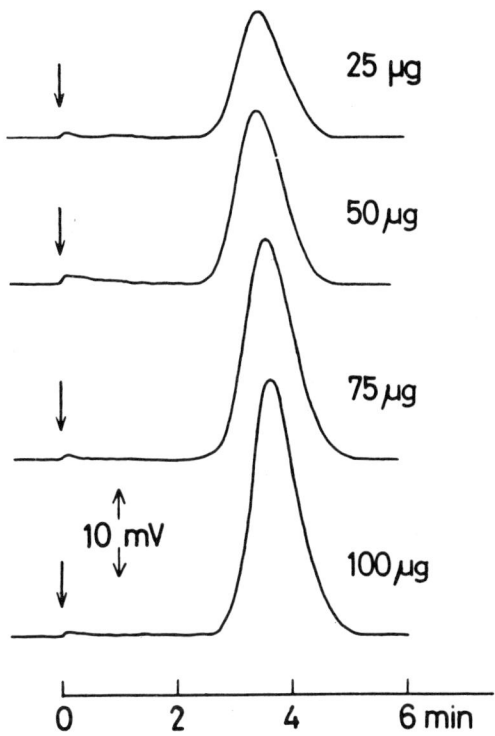

depends on temperature measurement careful thermostating of the total system would be necessary.

The Interferometer Detector

The principle of this detector is based on the change in effective path length for a beam of light passing through a cell when a solute is present in the mobile phase. If the light from the cell is focussed onto a photocell coincident with a reference light beam from the same light source interference fringes will be produced; the fringes will change as the path length of one light beam changes (from the detector cell) and thus as the concentration of solute increases in the detector cell a series of electrical pulses will be generated by the photocell as each fringe passes over it. The interferometer fringe pattern change is an indirect way of measuring the change in refractive index. In fact this detector is a form of refractometer detector: the pulses from the changing interference fringes also constitute an integration procedure the total number of pulses being directly related to the total mass of solute present in the eluted peak.

The effective optical pathlength, d, when the liquid-filled sample cell is placed in the interferometer depends on the change of refractive index of the liquid, Δn, and pathlength, l, of the cell according to the following equation

$$d = \Delta n l$$

Further it is possible to relate a number of fringes (N) (sensitivity), which move past a given point (or the number of cyclic changes of the central portion of the fringe pattern) to the refractive index change by the equation

$$N = 2\Delta n l / \Lambda$$

where Λ is the wavelength of the light used. The larger N is for a given Δn, the more sensitive the detector becomes. Therefore l must be large and Λ small as practical. The practical limit of the length of cell is determined by the maximum dead volume allowable for efficient flow through the cell and minimum dispersion.

A diagram of the optical system employed by the authors is shown in figure 5. Light from the source strikes a half silvered mirror M_1 and divides into two paths. Part of the beam is reflected onto a plain mirror M_2 and back through M_1 onto the photocell. P is a compensating glass plate. The other beam passes through the cell C to a plain mirror M_3 and thence back through the cell C to the half silvered mirror M_1 where it is also reflected onto the photocell. The trace representing a single peak for 8 µl of dioxane is shown in figure 6. Each peak in figure 6 represents the passage of a fringe across the photocell and the combination of the four peaks represents the single chromatographic peak. The number of fringe peaks will be directly proportional to the total refractive index change and this will be proportional to the total mass of solute present. Detection by this

Figure 5

The Optical System for the Interferometer Detector

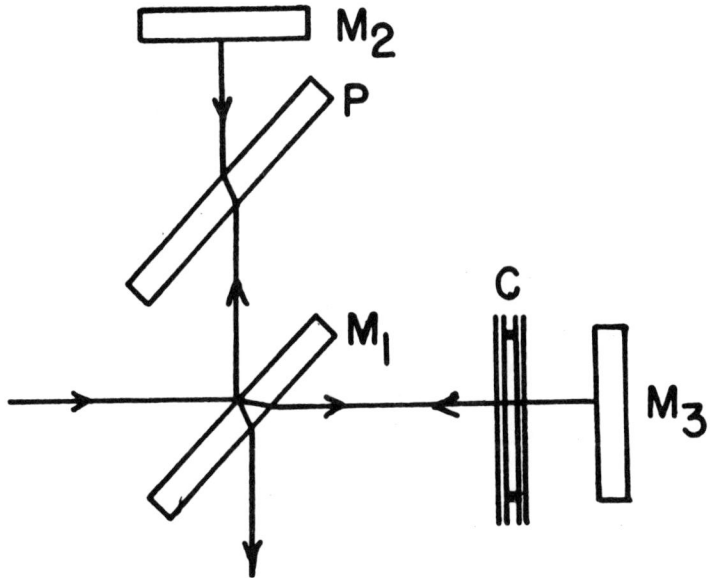

method of incompletely resolved peaks could be very confusing and difficult to interpret but this system is certainly one of the more novel methods for

Figure 6

An Elution Curve for Dioxane from an Interferometer Detector

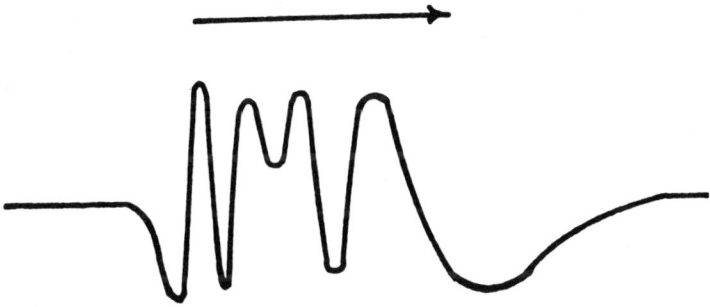

LC detection. In common with all bulk property detectors this detecting system is only applicable to isocratic and isothermal methods of development.

The Density Balance Detector

The density balance detector is a novel application of the Gow-Mac Gas Density Balance and has been described by Quillet(4). The gas density balance functions by measuring the differential flow across the base of two

Figure 7

The Gas Density Bridge Modified to Monitor Liquid Chromatography Eluents

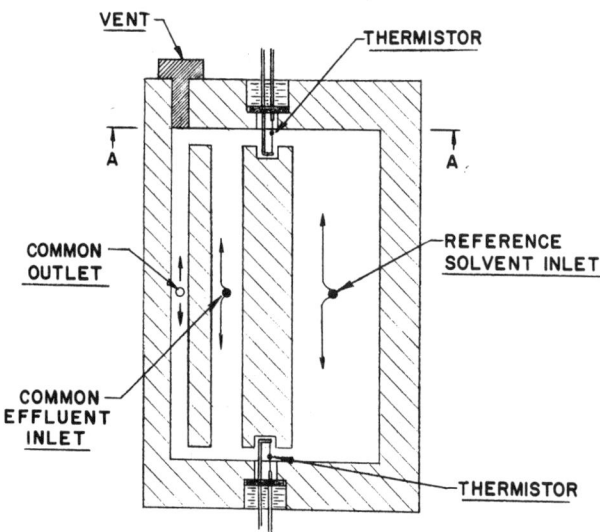

columns of gas one containing pure carrier gas the other containing the solute vapor as it is eluted from the column. The flow of gas resulting from the pressure difference at the base of each column is sensed by appropriately positioned thermistors. A diagram of a gas density bridge is shown in figure 7. Quillet situated the density balance in an appropriate thermostat and passed a reference flow of liquid through the reference cell and the column eluent through the sample cell. The normal ancillary

The Density Bridge Detector

Figure 8

Chromatogram from the Density Bridge Detector

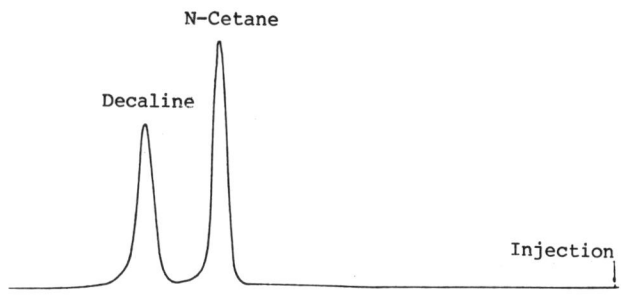

detector electronics were employed and a chromatogram obtained from his apparatus demonstrating the detection of 10 µl of cetane and decaline is shown in figure 8. According to Quillet the sensitivity of the detector was about 10^{-4}-10^{-5} g/ml and had a linear dynamic range of about one order. In the opinion of the author it is highly unlikely that the detector is responding to difference in density between the column flow and the reference flow but is more likely to be responding to changes in the thermal properties of the mobile phase such as thermal conductivity or specific heat. These properties of the mobile phase will change in the presence of the solute and could affect the output of the thermistor in much the same way as the thermal conductivity detector described earlier. The detector would have to be used under isocratic conditions of development but if investigated further might provide the basis of a very useful detector for preparative liquid chromatography where high sensitivities are not required.

The Vapor Pressure Detector

In the thermoelectric (vaporometric) determination of molecular weight a drop of a solution and a drop of solvent are suspended in an atmosphere saturated with solvent vapor. The steady state temperature difference which is developed between the drops is measured. For ideal solutions the temperature difference is linearly related to the solution strength,

$$\Delta T = d\, m,$$

where d is a constant (deg kg g-mole^{-1}); m is the solution molality (g-mole kg) and ΔT is a temperature difference between the drops (degrees). By

Figure 9

The Principle of the Thermistor Detector

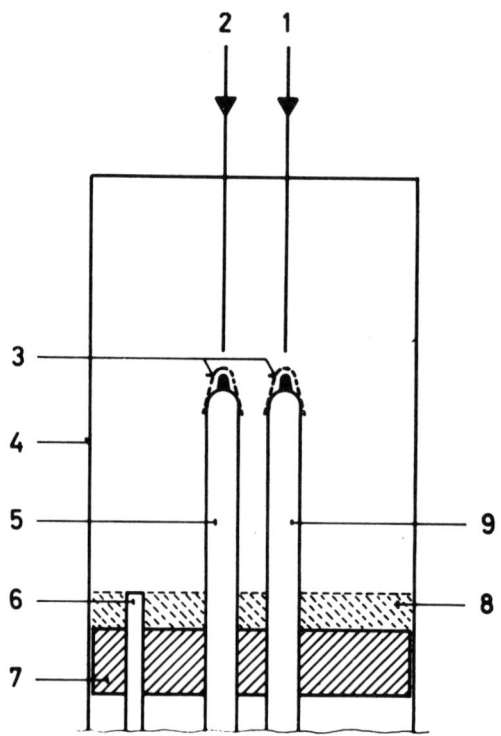

1, eluent inlet 2, reference inlet 3, gauge 4, copper chamber
5, solvent thermistor 6, overflow 7, Teflon plug
8, solvent pool 9, solvent thermistor

using thermistors to measure the temperature difference between the drops

$$\Delta R = -k\, m,$$

where k, calibration constant (Ohm kg g-mole^{-1}); ΔR, change in thermistor resistance in replacing solvent by solution on one sensor (Ohm). The principle has been employed by Simon et al. (5) to develop an effective liquid chromatography detector. A thermistor arrangement suited for use as a detector is shown schematically in figure 9. Such a detector is nonspecific for nonvolatile or slightly volatile solutes and gives a response proportional to the solute molarity. Pure solvent and column eluent are allowed to continuously drop onto each of two thermistor beads, capped with platinum gauze. The excess solvent and eluent flows over the outside of the thermistors into a reservoir at the base and then overflows to waste. The thermistors are enclosed in a well-thermostated copper cylinder to maintain strict temperature control. Each bead forms one arm of a Wheatstone bridge, and the "out-of-balance" signal is fed to a suitable recorder. The authors demonstrate that the time constant of the system can be made sufficiently small for use in liquid chromatography.

Poulson and Jensen (6) developed this detector still further and a diagram of their detector cell is shown in figure 10. The cell is symmetrical horizontally and contains the thermistor probes, T, near the center. The cell was machined from cylindrical aluminum stock to a finished size of 3-inch o.d. x 3.25 inches high and was machined flat on front and back to allow sealing of other members to the cell with Teflon gaskets. The front of the cell contained a 1.8-inch removable window (not shown) for access to the center of the cell in positioning components. To the rear of the cell was attached a 3-inch o.d. aluminum, solvent-venting chamber into which the wicks passed through ports, V. Solvent flowed from the wicks, W, into small basins which spilled over into a drain and was vented to waste outside the oven. The cell was bolted to a 3-inch aluminum pedestal, B, for additional mass. The void space in the cell was 2 inches high x 1.5 inches wide x 1 inch deep. The walls were lined with heavy blotter paper, while the top of the cell was closed by cover, C, and sealed with Teflon gasket, G. The whole cell was situated in a thermostated oven.

Fluids entered from the oven through 0.01-inch i.d. x 0.063-inch o.d. x 12-inch long stainless steel capillaries, I. The capillaries were mounted in tubing to 1/8-inch N.P.T. male adapters, F, drilled to clear the inlet

Figure 10

A Vapor Pressure Detector

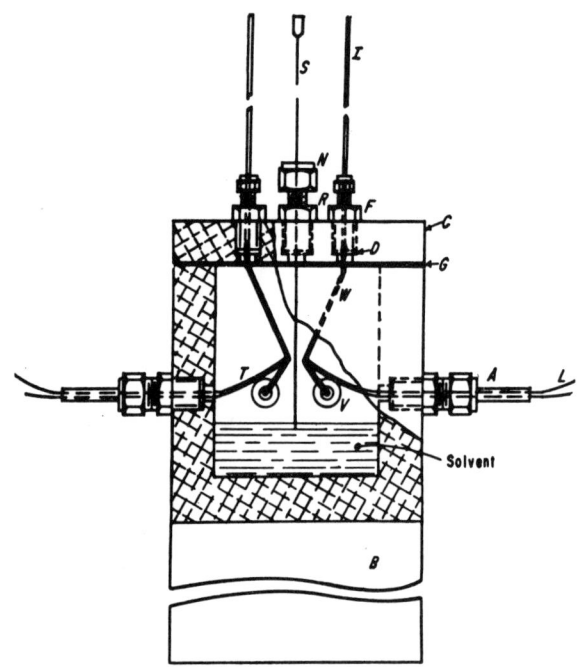

tubes. The fittings sealed against Teflon seals, D. The wicks fit snugly through holes in D and contacted the inlet tubes very near the surface of the seal. Teflon ferrule seals around the inlet tubes allowed positioning of the point of contact of tube and wick.

The level of the solvent reservoir in the cell was maintained through fill tube, S. The fill tube was a long hypodermic needle reaching outside the oven. The bottom of the fill tube was used to determine the solvent level. The tube fitting, through which this needle entered the cell, was sealed at the bottom by Teflon gasket, R, and also by a silicone rubber septum within nut, N.

The thermistors were sealed in the cell with tubing-to-1/8-inch N.P.T. male adapters, A, drilled to 0.080 inch. The Teflon-sheathed thermistors, although irregular in shape, sealed readily into the tubing adapters with standard 0.0625-inch i.d. Teflon ferrules which can be stretched over the probes. Thermistor leads, L, were taken out of the cell compartment with electrostatically shielded cable.

A chromatogram obtained from their apparatus is shown in figure 11. It is seen that reasonable sensitivity is achieved but the real attraction of this method of detection is that the response is proportional to the

Figure 11

Chromatogram for a Vapor Pressure Detector

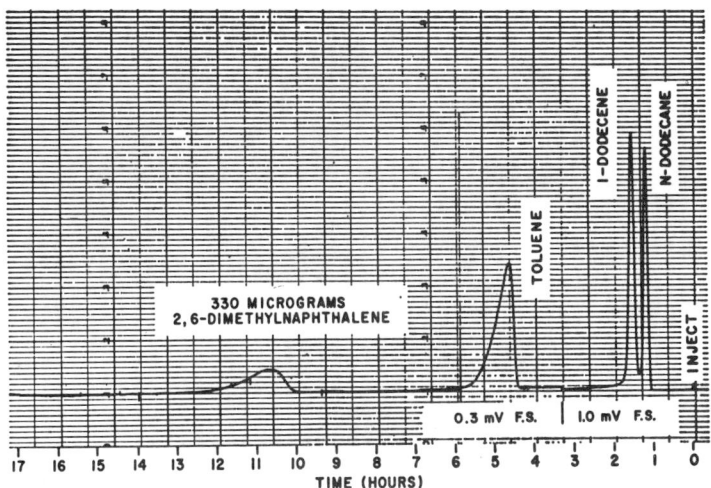

molecular weight of the solute and can be calibrated using a standard solute of known molecular weight. Further if the weight composition of a mixture is known from the results obtained using another detector the molecular weight of the unknown solutes can be calculated. For this reason it might well be worthwhile to develop this detecting system further and to use it in series with another detector. The resulting apparatus would provide gravimetric composition and solute molecular weight simultaneously.

REFERENCES

1. N. Fornstedt and J. Porath, J. Chromatogr., 42(1969) 376.
2. K. Ohzeki, T. Kambara and K. Saitoh, J. Chromatogr., 38 (1968) 393.
3. M. Bakken and V. J. Stenberg, J. Chromatogr. Sci., 9 (1971) 603.
4. R. Quillet, J. Chromatogr. Sci., 8 (1970) 405.
5. W. Simon, J. T. Clerc and R. E. Dohner, Microchem. J., 10 (1966) 445.
6. R. E. Poulson and H. B. Jensen, Anal. Chem., 40 (1969) 376.

PART 3

SOLUTE PROPERTY DETECTORS

CHAPTER 1

Principles of Detection

Solute property detectors function on the measurement of some property of the solute only, as opposed to bulk property detectors that measure some overall property of the column eluent which is modified by the presence of a solute. Solute property detectors are the most widely used type of detector in liquid chromatography. today. In this class of detectors are found the most sensitive, the most versatile and those with the widest linear dynamic range. However as already stated not all these attributes are contained in one single detector. Solute property detectors fall roughly into two broad classes, Specific Detectors and Transport Detectors.

Specific detectors monitor the total column eluent but measure some specific property of the solute that the mobile phase does not possess or has to a significantly lesser extent. It follows that only solutes having this specific property will be detected. Examples of specific solute property detectors are the UV detector and the Fluorescence detector. In this class of detectors are the most sensitive and those with the widest linear dynamic range. They are, however, not the most versatile as they restrict the choice of mobile phase to those solvents that do not possess the property which they measure. In practice specific detectors are often used in a compromise situation where they are acting as a hybrid between a solute property detector and a bulk property detector. An example of this is when the UV detector is employed with a mobile phase consisting of a small percentage of ethyl acetate in a hydrocarbon solvent. Ethyl acetate itself adsorbs UV light at the wavelength employed by UV detector and the effect of this adsorbance is eliminated by use of the same solvent mixture in the reference cell as will be discussed later. However as the mobile phase is now also providing a signal to the detector it is, in effect, partially a bulk property detector.

Transport detectors on the other hand are always truly solute property detectors. Transport detectors employ some transport medium such as a wire or chain to take a sample of eluent from the column, the solvent is then continuously removed by an evaporation procedure and the remaining solute exclusively monitored by combustion, pyrolysis or other suitable process. Transport detectors are the most versatile detectors and permit the use of any solvent or mixtures of solvents as the mobile phase providing they are reasonably volatile. Detectors of this type are thus very appropriate for use with gradient elution development. They are in general, however, less sensitive than the specific detectors and have, in fact, sensitivities more in common with bulk property detectors. Most transport detectors have a reasonably wide linear dynamic range.

Solute property detectors can place stringent demands on the purity of the solvent employed as the mobile phase. The UV detector often operates at sensitivities of the order of 10^{-8} or 10^{-9} g/ml and thus the mobile phase must be free of UV adsorbing materials down to this level. This is necessary because the impurities often have significantly different polarities to that of the bulk solvent and thus equilibrium with regard to the column stationary phase takes a considerable time to achieve. During the equilibrium time which may take many hours, the base line from the detector continuously drifts. In a similar manner, and for the same reason, trace high boiling contaminants in solvents will affect the stability of transport detectors as such impurities will not be removed in the evaporator and pass on to the detecting system.

In the author's experience all solvents employed for liquid chromatography should always be cleaned prior to use and this includes, spectra grade solvents and distilled in glass solvents. Solvents can be cleaned by passing them through packed beds of activated alumina or silica gel and subsequently filtering them through a 0.5 micron millipore filter; the latter procedure is particularly necessary when using the Waters 6000 pump. This pump is very sensitive to microparticulate matter in the mobile phase causing filter blockage and subsequent pump failure. When cleaning solvents by passage through a bed of adsorbent it is advisable to only pass three bed volumes of mobile phase through the system and then replace it with fresh adsorbent. If more solvent is passed through the bed, impurities possibly adsorbed in the media are liable to be eluted and recontaminate the purified solvent. It should be pointed out that cleaning solvents by adsorption will

Table I

Some Physical Properties of Solvents in Common Use in Liquid Chromatography

Solvent	Cut Off (nm)	Refractive Index	Density (20°C) g/ml	Boiling Point °C	Dielectric Constant
n-Pentane	205	1.358	0.6214	35.4	1.844
n-Heptane	197	1.388	0.6795	98.4	1.924
Cyclohexane	200	1.427	0.7739	80.7	2.023
Carbon Tetrachloride	265	1.466	1.5844	76.7	2.238
n-Butyl Chloride	220	1.402	0.8809	78.5	7.39
Chloroform	295	1.443	1.4799	61.1	4.806
Benzene	280	1.501	0.8737	80.1	2.284
Toluene	285	1.496	0.8623	110.6	2.379
Dichloromethane	232	1.424	1.3168	39.8	9.08
Tetrachloroethylene	280	1.938	1.3292	87.2	3.42
1,2-Dichloroethane	225	1.445	1.2458	83.5	10.65
2-Nitropropane	380	1.394	0.9829	120.3	25.52
Nitromethane	380	1.394	1.1313	101.2	35.87
n-Propyl Ether	200	1.381	0.7419	89.6	3.39
Ethyl Acetate	260	1.370	0.8946	77.1	6.02
Ether	215	1.353	0.7076	34.5	4.34
Methyl Acetate	260	1.362	0.9280	56.3	6.68
Acetone	330	1.359	0.7844	56.5	20.70
Tetrahydrofuran	225	1.408	0.8842	66.0	7.58
n-Propanol	205	1.380	0.7998	97.2	20.3
Ethanol	205	1.361	0.7850	78.3	24.0
Methanol	205	1.329	0.7866	64.7	33.6
Water	180	1.333	0.9971	100.0	80.3
Acetic Acid	210	1.329	1.0437	118.0	6.15

only remove impurities having significantly greater polarity than the bulk solvent. In most cases, however, the removal of such impurities provides solvents of adequate chromatographic purity.

The solvent or solvents used as the mobile phase have to be chosen to be compatible with the detector employed. In Table I the physical properties are given of a number of solvents commonly used in liquid chromatography. The solvents are arranged in order of their increasing polarity with respect to silica gel and the data provided can help the chromatographer in the choice of mobile phase to be used with any given detector.

Chapter 2

The Ultraviolet Absorption Detector

UV absorption detectors respond only to those substances that absorb ultraviolet light. A great many compounds fall into this category, including all substances having one or more double bonds (π electrons) and substances having unshared (non bonded) electrons, e.g., all olefins, all aromatics and compounds containing $>C=O$, $>C=S$, $-N=O$, $-N=N-$. The detector cell usually consists of a short length of tube, which is made to carry the eluent from the column and through which a beam of UV light is focussed onto a photoelectric cell. When solutes are present in the mobile phase, light is adsorbed and thus the intensity of light falling on the photocell is reduced producing an electrical output which can be amplified and fed to a recorder. The relationship between transmitted UV light through the cell and solute concentration is given by Beer's Law

$$I_T = I_o e^{-klc}$$

or $\text{Log } I_T = \text{Log } I_o - klc$

where I_o is the intensity of the light entering the cell
I_T is the light transmitted through the cell
l is the path length of the cell
c is the concentration of solute in the cell
and k is the extinction coefficient of the solute for the specific wavelength of the UV light

$$\text{thus } \frac{\partial (\text{Log } I_T)}{\partial c} = -kl.$$

It is seen that the sensitivity of the detector will be directly proportional to the value of the extinction coefficient (k) and the path length of the cell. It follows that to increase the sensitivity of the system l must be increased; however there is a limit to which the path length can be increased as discussed on page 25. To restrict band dispersion in the cell to a reasonable level the radius of the cell must be reduced as the path length is increased. This results in less light falling on the photocell which in turn reduces its signal to noise ratio. The two effects are in opposition and therefore this process of increasing sensitivity is limited unless photocells of significantly lower inherent noise levels are employed. Most UV cells have path lengths lying between 1 mm and 10 mm and diameters of about 1 mm.

The light source used in most UV detectors is a low pressure mercury vapor lamp. Such a lamp emits light at many discrete wavelengths the predominant line being at 254 nm. A filter is interposed between the lamp and the cell to remove the less predominant lines and provide virtually monochromatic light at 254 nm. Light of other wavelengths is often produced by allowing the monochromatic light at 254 nm to fall on a suitable phosphor and the light emitted by the phosphor used as the source for the detector cell.

UV cells can be very sensitive to both flow rate and temperature changes unless they are carefully thermostated. The effect of temperature control

Figure 1
A Thermostatted UV Cell

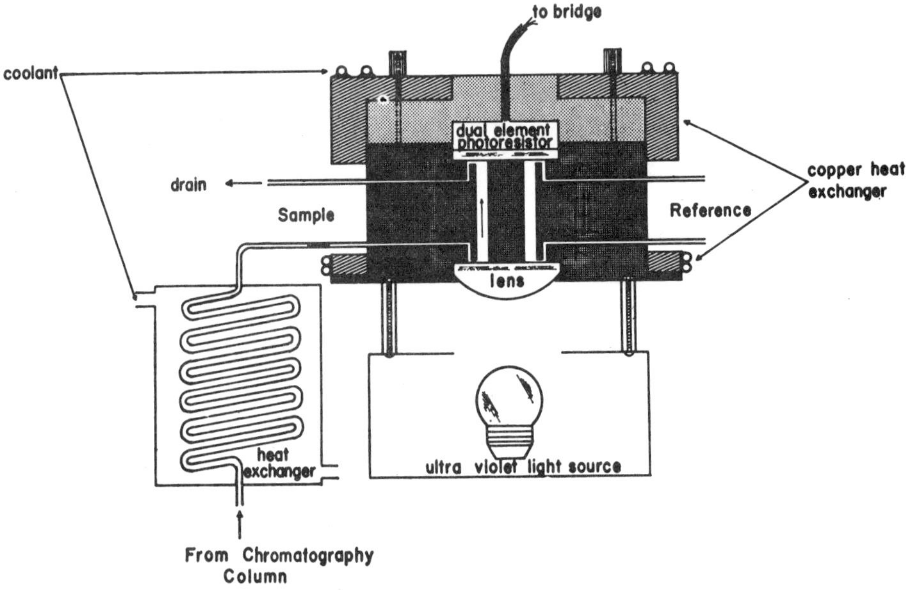

on the noise level of UV detectors was examined by Brooker (1). The modifications Brooker carried out are shown in figure 1. Essentially the eluent from the column is brought to the same temperature as the detector cell assembly by the use of an appropriate heat exchanger. The heat exchanger and the detector cell is thermostated by the same coolant stream. In the authors apparatus the heat exchanger consisted of a metal coil situated between the column and the detector and immersed in a coolant bath. If this apparatus were to be used with high efficiency columns the heat exchanger would have to be carefully designed so that the dispersion in the

Figure 2

Effect of Temperature Control on the Stability
of the UV Detector

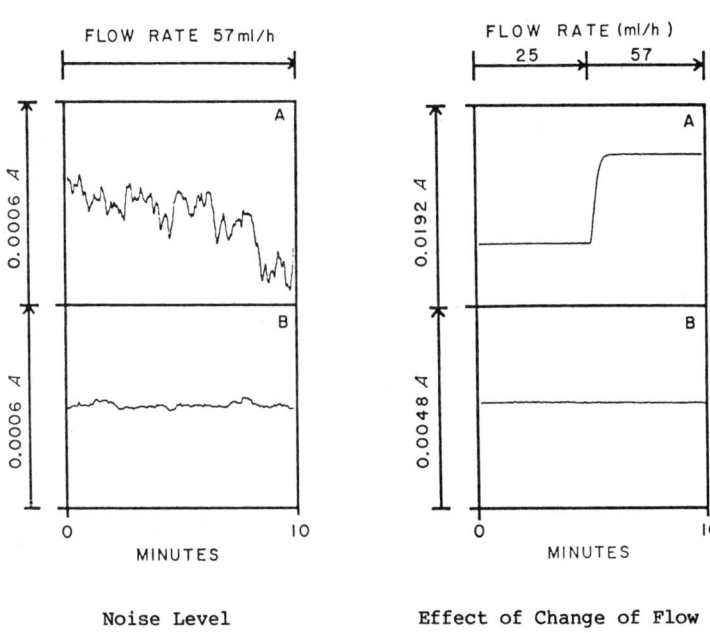

Noise Level   Effect of Change of Flow

A. No temperature control of cell or column eluent

B. Cell and column eluent measured at the same temperature

bath was kept to a level that would not impair the column performance. The effect of the thermostating system on detector stability is shown in figure 2. It is seen that the noise level of the thermostated system has been very significantly reduced and that the detector has become virtually insensitive to changes in flow rate.

From Beer's Law the concentration of solute in the cell is related to the logarithm of the intensity of the transmitted light. In the early models of UV detectors cadmium sulphide photocells were employed. These cells did not provide a response to the incident light intensity that could be described by a simple equation and so linearity between voltage output and solute concentration had to be obtained by employing an amplifier with a somewhat complex amplifying function. Modern detectors, however, use silicon photo diodes that provide a linear output with light intensity and thus by using a simple logarithmic amplifier the output can be related directly to solute concentration.

One of the first effective small volume UV detectors suitable for use with high efficiency columns was that described by Horvath and Lipsky in 1966 (2). In 1968 Kirkland (3) described in detail another small volume detector which subsequently formed the basis of the Dupont UV Detector. A diagram of the cell developed by Kirkland is shown in figure 3. The cell was cylindrical in shape, 1 mm in diameter and had a path length of 1 cm giving a

Figure 3

The Small Volume UV Cell Developed by Kirkland

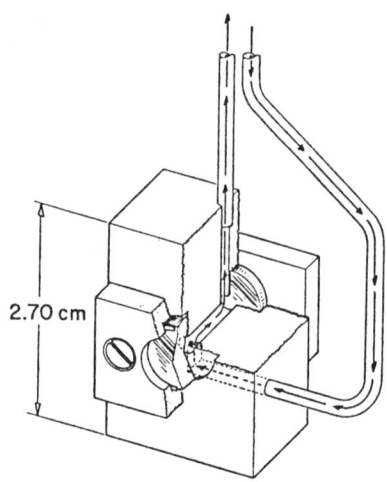

total cell volume 7.5 µl. Kirkland also used the 254 nm band from a mercury vapor lamp as the light source and claimed a full scale sensitivity of 0.01 absorbance unit and a noise level of ± 0.0002 absorbance. The linear dynamic range was 1.2-2 absorbance units equivalent to a concentration range of about 10^{-4} g/ml. Today most manufacturers have designed their own UV detectors which although each is novel in detail they all function on more or less the same principle as those described by Horvath et al. and Kirkland. Comparison between the use and performance of UV detectors and other types of detectors has been discussed by Polesuk (4) and Baker et al. (5) have compared the performance of different types of photometric detectors.

An example of a modern monochromatic photometric detector is the UV Monitor III manufactured by Laboratory Data Control. A photograph of this instrument is shown in figure 4. A totally removable cell permits the analyst to easily replace the standard 254 or 280 nm cell with his choice of a 350, 410, 440, 510 or 550 nm cell assembly. The cell is cylindrical in shape about 1.2 mm in diameter and has a path length of 1 cm giving a cell volume of 10 µl. The dimensions and volume of the connecting tube is not given. The cell is temperature compensated to minimize the effects of

Figure 4

The LDC UV Monitor III

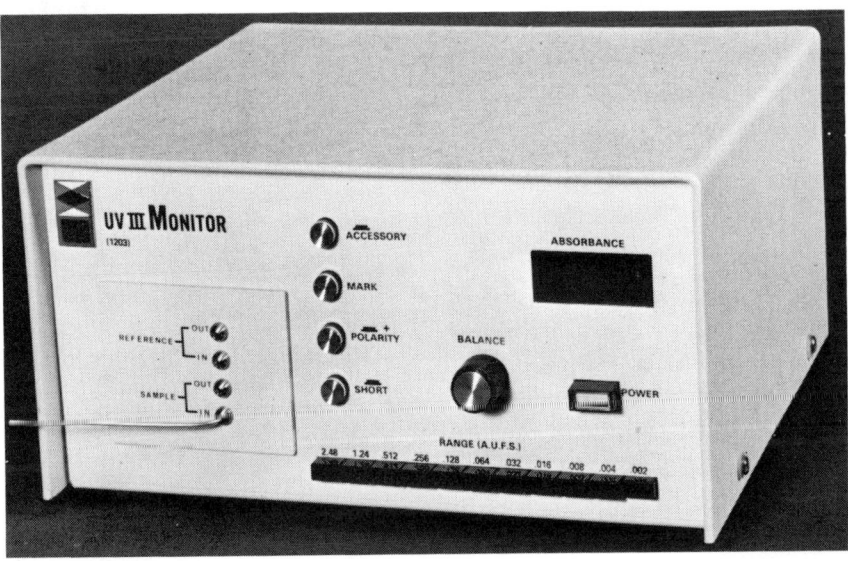

refractive index changes of the solvent. The optical assembly and electronic controls are contained in one unit thereby eliminating interconnecting cables. A slide out optical assembly makes it readily accessible for servicing and changing the cell.

Additional features of the UV Monitor III include ten push-button selectable sensitivity ranges from 0.002 to 2.56 A.U. digital display of differential absorbance at any wavelength, switch selectable time constants for 0.5 and 5 seconds, a polarity change switch, and event marker push button.

The performance data available for this detector is given in Table I. A more versatile instrument providing a wavelength coverage in the UV-VIS spectrum between 200 and 700 mm is the Spectromonitor I, also manufactured by

Figure 5

The LDC Spectromonitor I

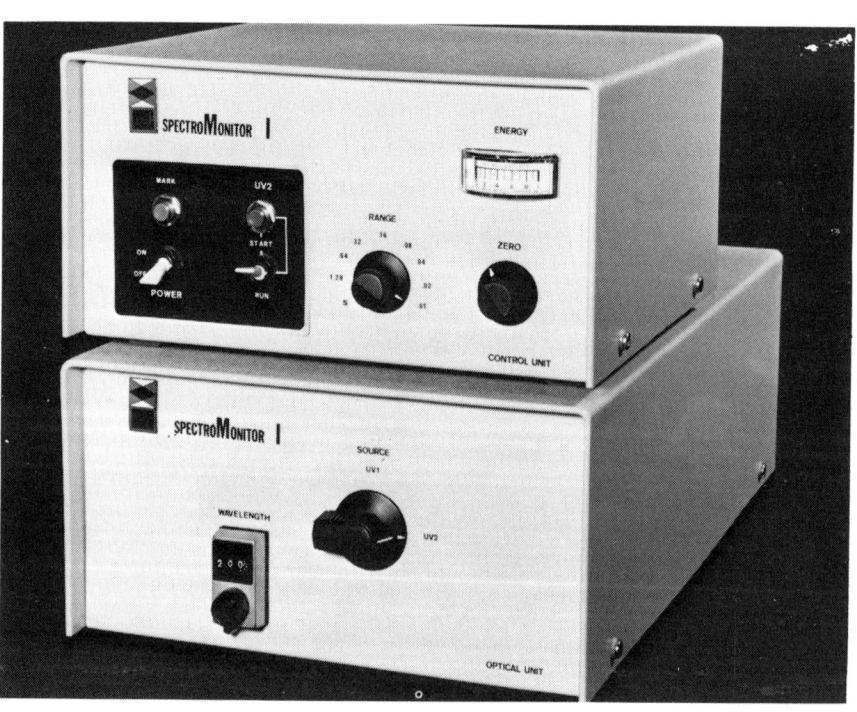

Table I

Specifications of the Laboratory Data Control UV Monitor III

Dynamic Range	2×10^{-5} - 2 AU
Response Index	Not given
Linear Dynamic Range	Quoted as the same as dynamic range
Detector Response	Not given
Noise Level	2×10^{-5} AU at 254 nm
Detector Sensitivity	1.6×10^{-8} g /ml of benzene @ 254 nm
Cell Shape	Cylindrical
Cell Dimension	Cell radius 1.2 mm Cell length 10 mm Cell volume 10 µl
Connecting Tube Dimension	Not given
Amplifier Time Content	Schedule 0.5 to 5 sec.
Wavelength Available	254, 280, 350, 410, 440, 510 and 550 nm

Laboratory Data Control. A photograph of the Spectromonitor I is shown in figure 5. The instrument consists of two separate units. The optical unit houses the cell, optics and photocell electronics and can be located conveniently near the solvent stream to be analyzed. The control unit contains the power supplies and output controls and a socket for interconnection with the optical unit by an appropriate cable.

The sample stream and a reference stream are connected to the measurement cell at the inlet tubes on the left side of the optical unit and separate outlet tubes above their respective inlets are provided for further processing or sample collection. An output signal of 10 millivolts is provided for recording absorbance and an energy meter on the front panel aids in selecting the source switch position.

A convenient front panel switch selects the appropriate source for the wavelength range desired.

Wavelength (nm)	Source
390-700	Tungsten
310-400	Tungsten
200-350	Deuterium

The cell has a relatively low dead volume and compensation for solvent absorbance and source energy variation is made by using a dual beam (sample-reference) optical system.

Minimum spacing of sample and reference beams is achieved by a very small quartz fiber optic beam splitter allowing the use of a single quartz

Figure 6

The Optical System of the Spectromonitor I

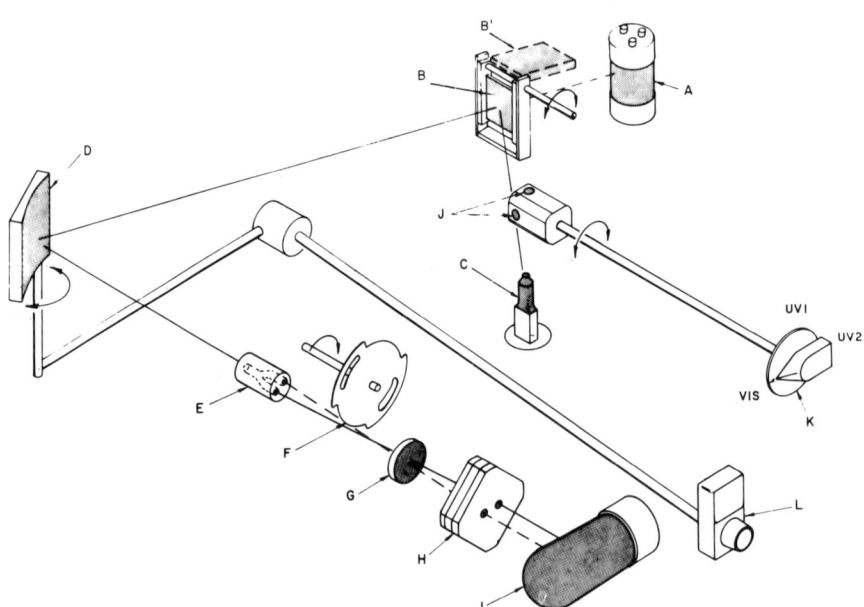

lens which focuses both beams on their respective cell paths.

Wavelength dispersion is by means of a concave holographic diffraction grating. The concave nature of the grating focuses the entrace slit on the exit slit without the usual two concave mirrors. A diagram of the optical system is shown in figure 6.

For UV measurements a continuum of energy as supplied by the deuterium source lamp (A) is directed toward the diffraction grating (D). By rotation of the grating to the appropriate angle of incidence to this beam of energy the desired wavelength will be projected into the beam splitter (E).

Two separate, equal intensity beams produced by the beam splitter are focussed onto the dual flow cell (H) by a single quartz lens (G). The sample and reference cavities of the cell are alternatively illuminated by interposing a rotating disc optical chopper (F). Energy which passes through the sample fluid in the cell to the photomultiplier detector (I) is compared to the energy passed through the corresponding reference path electronically.

The operation is the same for visible and near UV wavelength absorbance measurements except that the source change mirror (B) is positioned in the entrance beam to direct the energy from the tungsten source lamp (C) toward the diffraction grating. The geometric configuration is arranged such that the virtual image of the tungsten lamp is at the same position as the deuterium lamp. This helps to provide wavelength accuracy over the entire wavelength range with only single point calibration. The appropriate filter (J) is selected by the front panel control which also moves the source change mirror (B) and selects the source (B or B') to be energized. A UV blocking filter (J) is used to prevent second-order dispersion wavelengths from interfering with when operating in the 390-700 nm range. A visible radiation blocking filter (J) is used to minimize stray light when operating in the 310-400 nm range.

A photomultiplier detector with broad spectral response is used to convert the light intensity to an electrical signal, the output of which is amplified and fed to a logarithmic converter to ensure that the signal is directly related to solute concentration.

The useable energy varies as a function of source output spectrum, source lamp aging, detector spectral response, and reference stream absorbance. To, compensate for this an automatic gain control system is employed which appropriately adjusts the photomultiplier output.

Since the deuterium lamp is basically a gas discharge tube, a stable current supply is required for operation and a special circuit is used that provides a constant current to the tube and compensates for fluctuations in line voltage. The performance data available for the UV Monitor I is given in Table II and an example of a chromatogram obtained from this instrument is shown in figure 7. A continuously variable wavelength source can be extremely useful in practice. The detector sensitivity can be enhanced for any particular solute by choosing the wavelength at which the solute exhibits maximum adsorption. By this method very small traces of a compound can be detected and in some cases the detector can be made almost solute

Table II

Specifications of the Laboratory Data Control UV Monitor II

Dynamic Range	2×10^{-4} - 4.28 AU
Response Index	Not given
Linear Dynamic Range	Quoted as the same as the dynamic range
Detector Response	Not given
Noise Level	2×10^{-4} AU @ 254 nm 3×10^{-4} AU @ 210 nm
Detector Sensitivity	1.6×10^{-7} g/ml of benzene @ 254 nm
Cell Shape	Cylindrical
Cell Dimensions	Cell length 10 mm Cell diameter 1.3 mm Cell volume 13
Connecting Tube Dimension	Not given
Amplifier Time Constant	Selectable 1 sec to 5 sec
Wavelength Available	200-700 nm continuously variable

specific. The continuously variable wavelength source can also be used in the converse manner. If there is a major peak that obscures the presence of another solute due to column overload, a source wavelength can be chosen at which the major component exhibits a minimum adsorption. This reduces the detection sensitivity specifically for that peak and thus permits the other component to be more clearly identified. Employing source wavelengths far from those at which the whole sample adsorbs can also reduce the sensitivity of the detector for all solutes and allow the detector to be used for preparative chromatography where the large sample sizes employed can saturate the detecting system.

The UV detector is probably the most useful and extensively used detector. If designed correctly it has the highest sensitivity for general detection purposes, the widest linear dynamic range and sufficient versatility to permit its use with a reasonably wide range of solvents. It is basically a fairly simple detector and generally not expensive. The UV detector can be used with integrating systems and on line with computers and provided the dead volume is relatively small it can be effectively used with

Figure 7

Chromatogram Obtained from the LDC Variable
Wavelength Detector

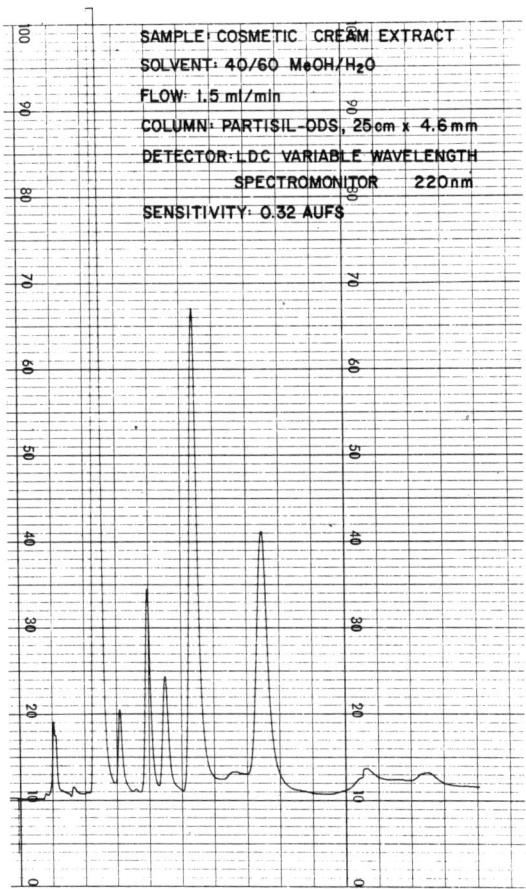

modern high efficiency columns. In fact the dead volume of the detector could be made significantly smaller than those presently available.

REFERENCES

1. G. Brooker, Anal. Chem., 43 (1971) 1095.
2. C. G. Horvath and S. R. Lipsky, Nature, 211 (1966) 748.
3. J. J. Kirkland, Anal. Chem., 40 (1968) 391.
4. J. Polesuk, American Laboratory, May (1970) p. 27.
5. D. R. Baker, R. C. Williams and J. C. Steiden, J. Chromatog. Sci., 12 (1974) 444.

Chapter 3

The Fluorometric Detector

Fluorescence is a specific type of luminescence. When molecules are excited by electromagnetic radiation to produce luminescence this phenomena is termed photoluminescence. If the release of electromagnetic energy is immediate or stops on the removal of the exciting radiation the substance is said to be fluorescent. If, however, the release of energy is delayed or persists after the removal of the existing radiation then the substance is said to be phosphorescent. Due to its persistence phosphorescence is of little use as a process for liquid chromatography detection. Fluorescence, however, has been shown to be extremely effective and detectors based on fluorescence measurement have provided the highest sensititivities available.

When light is adsorbed by a molecule a transition to a higher electronic state takes place and this absorption is highly specific to the molecule concerned; radiation of a particular energy or wavelength is only adsorbed by a specific molecular structure. If electrons are raised, due to absorption of energy, to an upper excited singlet state then such transactions are responsible for the characteristic visible or ultraviolet absorption spectra observed for such compounds. If the excess energy is not dissipated rapidly by collisions with other molecules or by other means the electron will return to the ground state with the emission of energy in the form of electromagnetic radiation. This effect is called fluorescence. As some energy is always lost before emission occurs the emitted fluorescent energy is always of longer wavelength than the absorbed or exciting radiation. Excellent discussions on the theoretical basis of fluorescence have been given by Guilbault (1) and Udenfriend (2) and readers interested in learning more of the theory and use of fluorescence are recommended to refer to these two books.

In comparison with other detection techniques fluorescence affords greater sensitivity to sample concentration but less sensitivity to instrument instability and such macroscopic properties as temperature and pressure. In part this may be attributed to the nature of the measured experimental parameter, which in fluorescence detection is a signal superimposed upon a very low background whereas in, for example, optical absorbence measurements, the signal is superimposed upon a high and sometimes unstable background. The major disadvantage of fluorescence detection is that not all compounds fluoresce under normal HPLC conditions. However, the large number of fluorescent materials, including biochemicals, foods, drugs, dye intermediates, etc. recommend the detection technique to specific areas of application. Further the scope of fluorescence detection can be extended by the use of fluorescent reagents such as Fluoropa (o-

Figure 1

The LDC Fluoromonitor

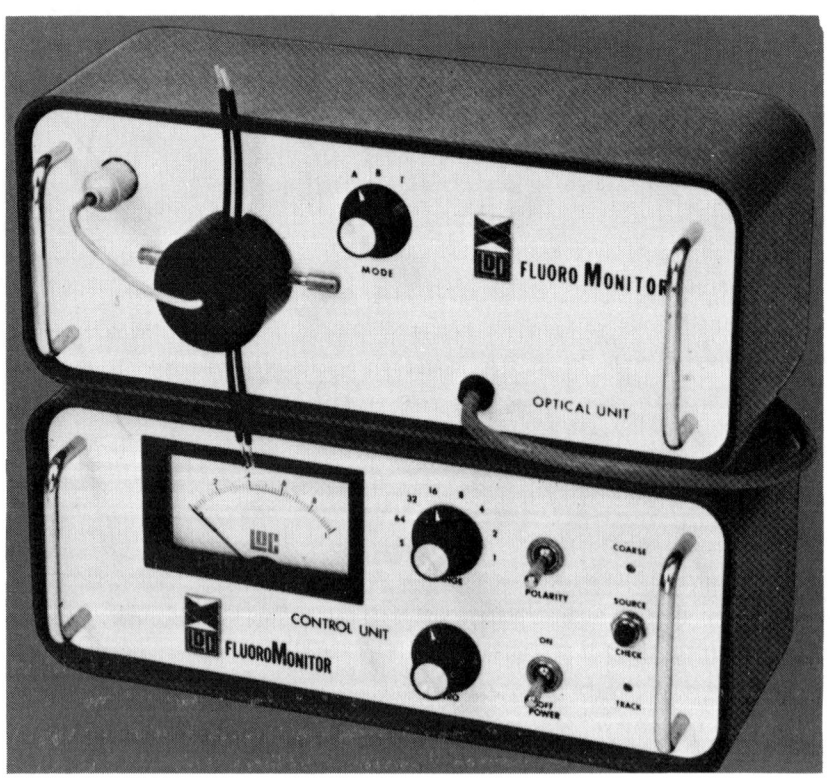

phthalaldehyde manufactured by Durrum) and Fluorescamine (4-phenylspiro (furan, 2H-1-phthalan) 3',3'-dione manufactured by Hoffmann-La Roche). These reagents react with primary amines such as amino acids to produce derivatives that fluoresce when irradiated with UV light. The use of such reagents requires a post column reactor of the correct design to restrict band dispersion and loss of resolution as discussed in Part 1, Chapter 5.

There are a number of fluorescent detectors commercially available all of which are effective but the majority of which have been designed for use with the large bore low efficiency ion exchange columns and thus have large cell volumes. Large cell volumes, many in excess of 25 µl, make them unsatisfactory for use with high efficiency columns packed with microparticulate bonded ion exchange packings presently available. One of the first fluorescence detectors that was commercially available with relatively small cell volume was the Fluoromonitor manufactured by Laboratory Data Control. This detector has a dead volume of 13 µl and although it is likely to be superceded in the near future by a more versatile detector having a dead volume of 8 µl, it is one of the less expensive fluoromonitor detectors and as it will be available for some time to come its construction

Figure 2

The Fluoromonitor Cell Assembly Optical System

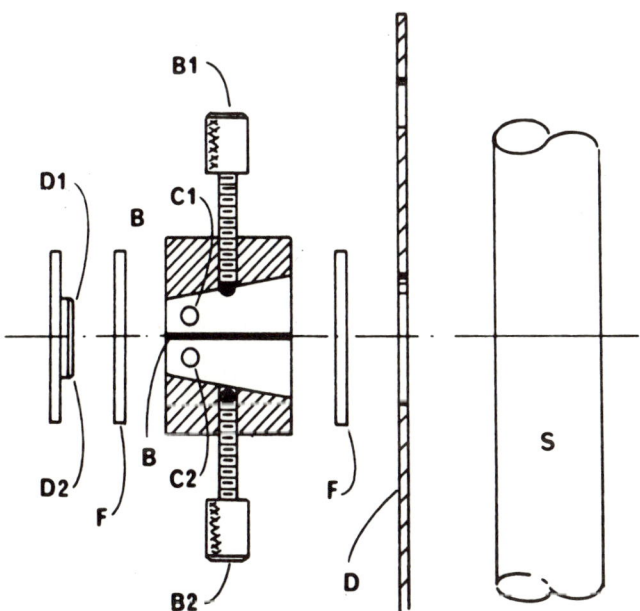

and performance will be described. A photograph of the LDC Fluoromonitor is shown in figure 1.

The Fluoromonitor is of modular design consisting of an electronic control unit and an optical unit. The optical unit contains the near UV excitation source lamp, sample and reference flow cells, and the solid state detector photocell. The control unit is cable-connected to the optical unit and contains the polarity switch, the sensitivity switch and the zero control.

The cell assembly of the Fluoromonitor is shown in figure 2. The excitation lamp S is a low pressure, hot cathode mercury lamp with a phosphor coating that emits near UV light peaking around 360 nm. Visible light is blocked by filter F before entering the large diameter end of the

Table I

Specifications of the Laboratory Data Control Fluoromonitor

Dynamic Range	Ca. 10^4
Response Index	Not given
Linear Dynamic Range	Varies with solute detected
Detector Response	Not given
Noise Level	Not given
Sensitivity	10^{-9} g/ml of quinine sulphate
Cell Shape	Cylindrical
Cell Dimension	Length - not given Radius - not given Volume - 13 μl
Connecting Tube	Dimension not given
Amplifier Time Constant	Not given
Wavelength of Excitation Light	360 nm
Wavelength of Emission Light	400-700 nm

cone condenser B. The highly reflecting internal surfaces of the cone condenser and its metal bifurcating plate direct the excitation light to cell chambers C1 and C2. A fluorescent coating on two adjusting thumb screws, B1 and B2, is provided to compensate for background solvent fluorescence. Emission of visible light from cell chambers C1 and C2 passes through a sharp cut-off UV blocking filter F and impinges on the corresponding photosensitive elements of the dual photocell D_1 and D_2.

The electronic measuring system is coupled to the dual detector in a Wheatstone bridge circuit activated by a feedback power supply which not only measures the difference between the photoconductor elements, but also compensates for fluctuations in light intensity. When a fluorescent solute band passes through one of the flow cells, the amount of visible light

Figure 3

A Chromatogram of a Standard Amino Acid Mixture from the Fluoromonitor

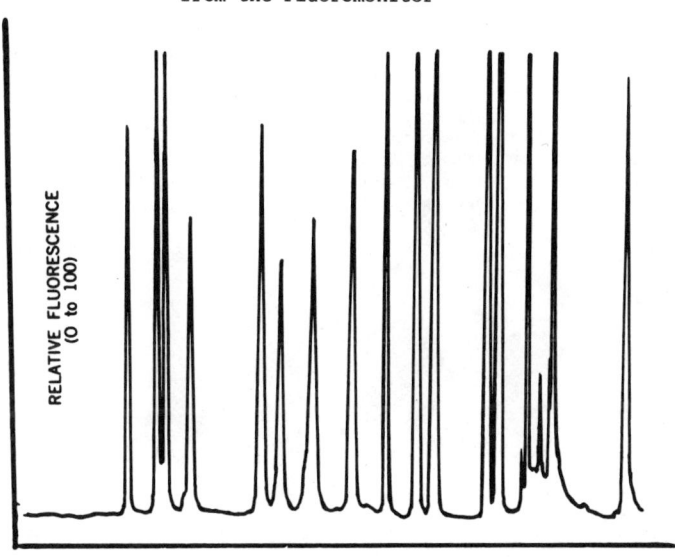

reaching the corresponding photocell element is increased. This unbalances the bridge, creating an output voltage which then is fed to the recorder. The specifications available for this detector are shown in Table I and an example of a chromatogram of a standard amino acid mixture is shown in figure 3. The chromatogram was obtained using the fluorescamine reagent in conjunction with a post column reactor and the sensitivity realized was about 10^{-10} g/ml.

For qualitative analytical work fluorescence detection may be used almost without restriction. However, for quantitative analysis caution is emphasized because of the non-linear relation between sample size and fluorescence intensity. The non-linearity arises from the unequal distri-

Figure 4

The Du Pont Model 836 Fluorescence/Absorbance Detector

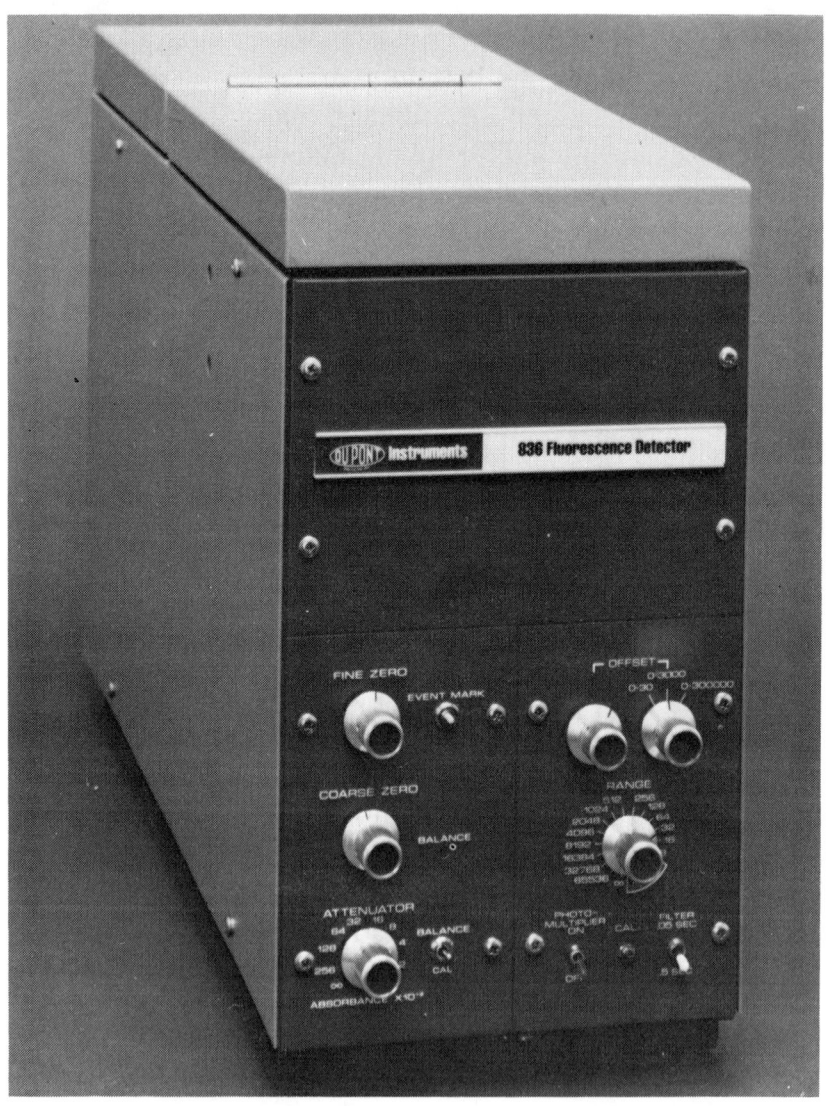

bution of light in the sample detection region. For this reason careful calibration should be carried out over the range of solute concentrations to be determined to ensure the response is linear over the range of interest. Alternatively appropriate correction factors can be determined.

The combination of UV adsorption and fluorescence becomes a very powerful detection technique realizing the advantages of both systems simultaneously. Further as this can be achieved using a simple detector cell, dual detection can be obtained with minimum dispersion of the solute band. A dual UV fluorescence detector has been described by Steichen (3) which became the basis of the Du Pont Model 836 Fluorescence/Absorbance Detector. A photograph of the instrument is shown in figure 4. The absorbance unit is available with a low pressure mercury source to provide UV light at 254 nm, a medium pressure lamp with appropriate filters to provide light at wavelengths of 254, 280, 313, 334 and 305 nm and a quartz iodine source providing light at wavelength 380-650 nm in 10 nm increments. As this instrument employs filters and does not utilize phosphors to generate additional wavelengths in the UV region, noise levels are minimal and energy output is optimized. The electronics utilize temperature-stabilized solid state circuits.

A diagram of the optical system is shown in figure 5. Light from a lamp is focussed onto a partially silvered mirror, 10% of the light being

Figure 5

The Optical System of the Du Pont 836
Fluorescence/Absorbance Detector

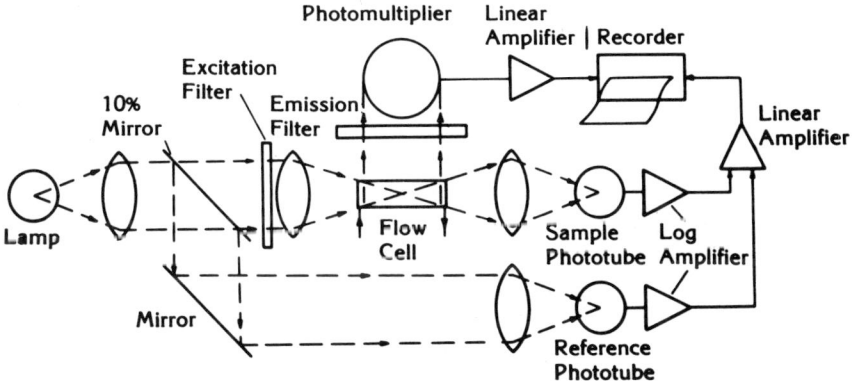

reflected via another mirror through a lens onto the reference photoelectric cell which compensates for changes in light intensity from the lamp. The remainder of the incident light falls onto an excitation filter and the selected wavelength passes through the filter and thence through the

Table II

Specifications of the Du Pont Model 836 Fluorescence/Absorbance Detector

Dynamic Range	Not given estimated 10^4
Response Index	Not given
Linear Response Range	Various with solute detected
Detector Response	Not given
Noise Level	Not given
Sensitivity	Ca 10^{-9}-15^{10} g/ml
Cell Shape	Cylindrical
Cell Dimension	Length 20 mm Diameter 1 mm Volume 16 µl
Connecting Tube	Dimension not given
Amplifier Time Constant	Not given
Wavelength of Excitation Light	325-385 nm
Wavelength of Emission Light Detector	451 nm

sample cell: the transmitted light then passes via another lens onto the sample photoelectric cell. The output from the two photoelectric cells each pass to a logarithmic amplifier and the output from each amplifier to a difference linear amplifier and thence to one channel of a two pen recorder. Mounted horizontal to the cell is an emission filter that eliminates the incident light and the transmitted fluorescent light falls onto a photomultiplier. The output from the photomultiplier then passes to another linear amplifier the output from which passes to the second channel of the two pen recorder. In this way simultaneous absorbance and fluorescence recordings of the column eluent are obtained. The specifications available for this detector are shown in Table II.

An example of a dual trace recording of a chromatogram of peanut butter extract is shown in figure 6. It is seen that the two compounds aflatoxin G_1 and aflatoxin G_2 at a level of 1 ppb are clearly shown on the fluorescence

Figure 6

A Dual Trace Recording of Peanut Butter Extract from the Du Pont 836 Fluorescence/Absorbance Detector

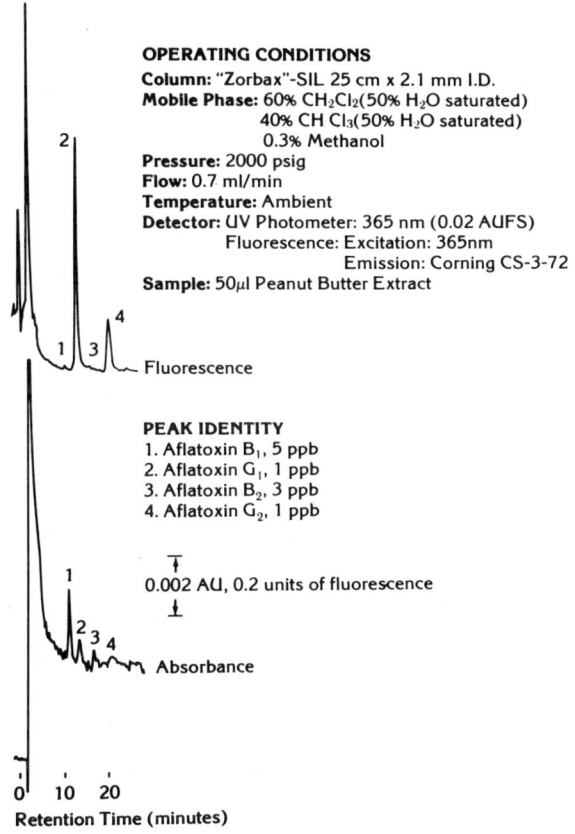

trace but are hardly discernible on the absorption trace. Conversely aflatoxin B_1 and aflatoxin B_2 at levels of 5 and 3 ppb respectively are clearly shown on the absorption trace but are not discernible from the noise on the fluorescence trace. The chromatograms clearly demonstrate the advantages of employing the dual detecting systems simultaneously.

The fluorescence detector has the highest sensitivity of all detectors presently developed. It is, however, a very specific detector and thus limits its field of application. Providing pure solvents, free of fluorescing substances, are employed it can be effectively used with gradient elution. However, it is often difficult to remove all traces of fluorescent materials from some commonly used solvents. The dynamic range is fairly large (10^4) but its linear dynamic range may be restricted for certain solutes to only one order. In general the response of the detector tends to be somewhat non linear, due partly to the nature of the formation of the fluorescence, partly due to the effect of quenching that can occur with high solute concentrations and partly due to the effect of the adsorption of incident light by other solutes being present or the solvent itself. Under some conditions, at high sensitivity, linearity can be assumed and accurate quantitative results obtained. However, it is recommended that in all quantitative analysis the detector linear dynamic range should be determined over the solute concentration range of interest. The fluorescence detector is used to its best advantage when employed in conjunction with an adsorption detector to provide simultaneous output: it is not recommended for general use but for specific applications.

REFERENCES

1. G. G. Guilbault, Practical Fluorescence , Marcel Dekker, Inc., New York, 1973.
2. S. Udenfriend, Fluorescence Assay in Biology and Medicine , Academic Press, New York, London, 1962.
3. J. C. Steichen, J. Chromatog., 104 (1975) 39.

Chapter 4

The Polarographic Detector

Polarography is an electrochemical process used in chemical analysis where a solution is electrolyzed at a small, polarizable dropping mercury electrode in conjunction with a large non polarizable electrode. If the solution contains substances that will oxidize or reduce a stepwise increase in the current results from a progressively increased applied voltage, the potential at which the step occurs is a characteristic of the solute. The first worker to employ polarography as a detecting system in liquid chromatography was Vermula in 1952 (1) and since that time many workers have explored the technique as a detection method.

Polarography using the dropping mercury electrode has some advantages as a liquid chromatograph detecting system. The mercury drops result in a continuously new and clean electrode surface being exposed to the solution, however, the inherent mechanical limitation make it, overall, an unattractive detecting system. The dropping electrode in the continuous operation has also a limited anodic potential range which precludes its use in the electro oxidation mode and this confines its detection capabilities to electro reducible compounds.

Solid electrode systems offer far more promise for electrochemical detectors providing the surface is unadsorptive, is electrochemically reproducible and does not readily oxidize. Joynes and Maggs (2) investigated a commercially available carbon-impregnated silicone rubber membrane electrode, however, in Joynes' system the electrode surface was not readily reconditional. MacDonald and Duke (3) and Adams (4) have described the design and performance of carbon paste electrodes and Takata and Muto (5) have investigated the characteristics of carbon, platinum and silver gauze. A number of other solid electrode materials have been examined (6, 7), but the maintenance of constant surface activity is still a problem. The stability of solid electrodes still varies due to adsorption effects, formation of adhered film and these factors contribute significantly to short and long term noise. Attempts have been made to develop cleaning procedures the most common being mechanical abrasion and potential cycling to dissolve any adhered films but no general satisfactory method has been developed. Work on solid surface electrode materials so far seem to indicate that glassy carbon or graphite inpregnated styrene are likely to be the most useful.

Most modern electroanalytical techniques employ a 3-electrode cell arrangement where the conventional working and reference electrodes are supplemented by a third counter electrode. In the normal potentiostatic mode the reference electrode is used solely to monitor the potential of the working electrode and any deviation from the preset voltage causes current feedback via the counter electrode to restore the balance. For ideal cell geometry the working and reference electrode should be located as close together as possible and out of the current path between working and counter electrodes. This optimum arrangement may present a problem in trace analysis because of leakage of reference electrolyte into the test solution. With continuous flow cells, however, the situation is much more favorable since both the reference and counter electrodes can be located downstream of the working electrode; thus leakage of reference electrolyte or evolution of oxygen from the counter electrode does not cause any interference.

An example of a detector cell designed by Fleet and Little (8) utilizing the dropping electrode is shown in figure 1. The use of a conically ground dropping mercury electrode (DME) enabled the volume of the cell to be maintained between 2-5 µl and at the same time provided a larger working surface at the inlet stream of mobile phase from the column. The conical DME also provides a more uniform electric field and prevents phase change effects which are observed with a conventional capillary when used in the a.c. mode.

The reference electrode consists of a silver/silver chloride element connected to the cell chamber via a Vycor porous glass frit. A platinum or titanium tube served both as exit line and counter electrode.

Another detector cell system developed by Fleet and Little (8) based on the work of Yamada and Matsuda (9) and termed the Wall-Jet Electrode Detector (WJED) is shown in figure 2. In this design the inlet solution is introduced via a nozzle (1) and impinges normally on the planar disc electrode (2). The solution then exits to waste via two diametrically opposed channels (4) and (5). A silver/silver chloride reference electrode is connected to channel (4) via a Vycor porous glass plug while the exit line (5) consists of a 1.5 mm platinum tube which also serves as a counter

Figure 1

The Detector Cell Designed by Fleet and Little
1, conical DME; 2, column inlet; 3, reference electrode; 4, counter electrode.

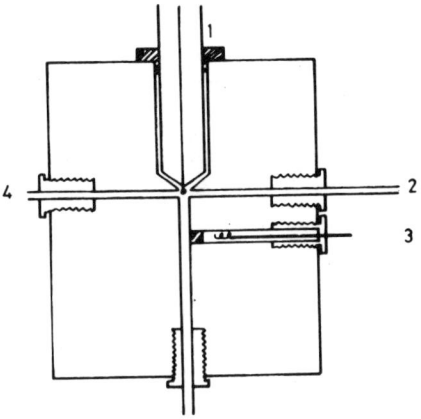

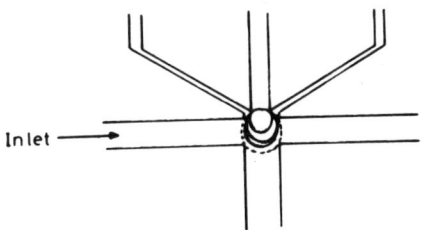

electrode. The body of the disc electrode (3) is threaded so that it is possible to control the distance (α) between the nozzle tip and the surface of the electrode. Thus the effective cell volume was adjustable and could be reduced to 0.5 μl.

The wall-jet detector has some useful characteristics. The rapid convective mass transfer which takes place at the point of contact results

Figure 2

The Wall-Jet Electrode System

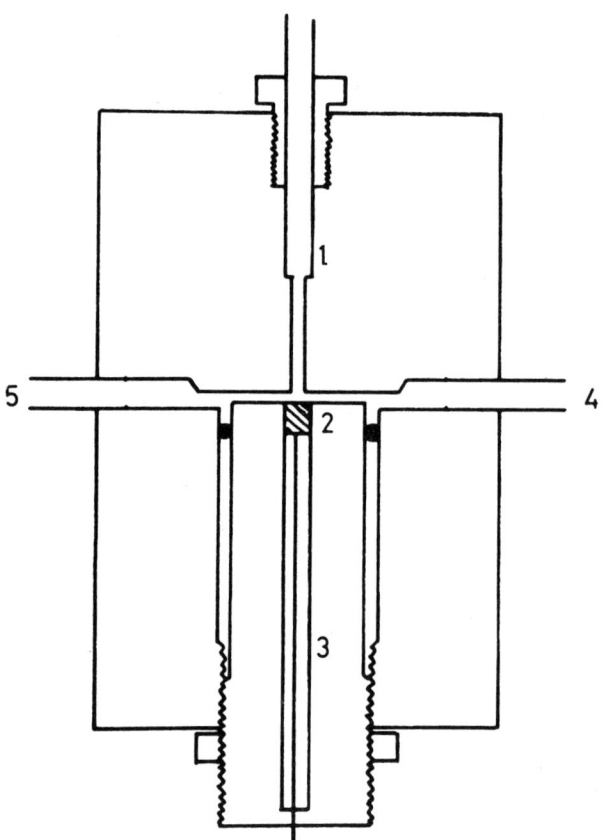

in very high sensitivity. Initial calibration using ferrocyanide as electroactive species gave a detection limit of ca. 10 ng. The glassy carbon working electrode has a wide operating range, from -1.3 V on the cathodic side to +1.5 V in the anodic region. Consequently the system can be operated both in the electro-reduction and electro-oxidation modes.

Finally studies indicated that this arrangement of electrode is sensibly free from problems of non-reproducibility caused by surface adsorption

Figure 3

Effect of Working Electrode Potential on Detector Response

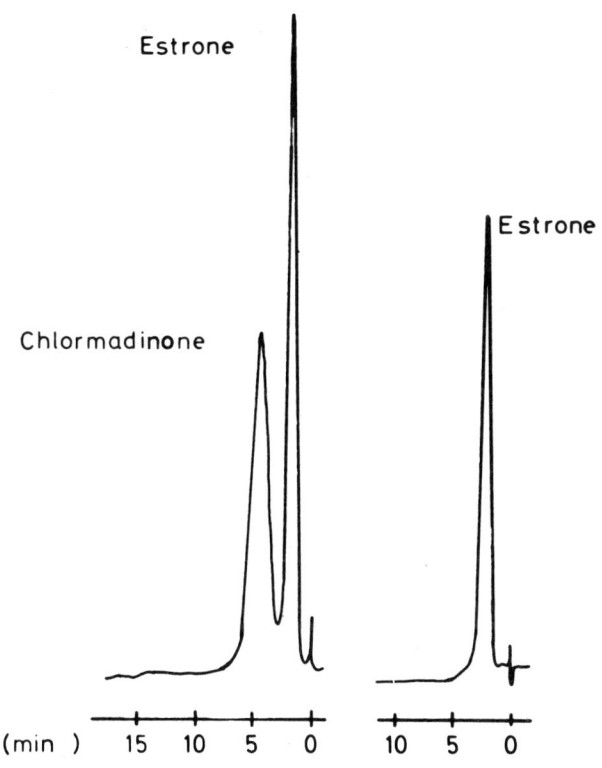

effects. It is very probable that the electrode is "cleaned" of adsorbed electrolysis products by the washing effect of the rapid incoming solution.

Figure 3 illustrates the use of control of working electrode potential to influence the selectivity of the detector. The chromatograms shown are for a mixture of two steroids chloromadinone, a keto steroid and estrone, a hydroxy steroid using the WJED and for comparison, a UV detector at a fixed wavelength of 285 nm. By controlling the WJED working potential at +0.8 V

136　Liquid Chromatography Detectors

Figure 4

Chromatogram of Amino Acids from a Polarographic Detector

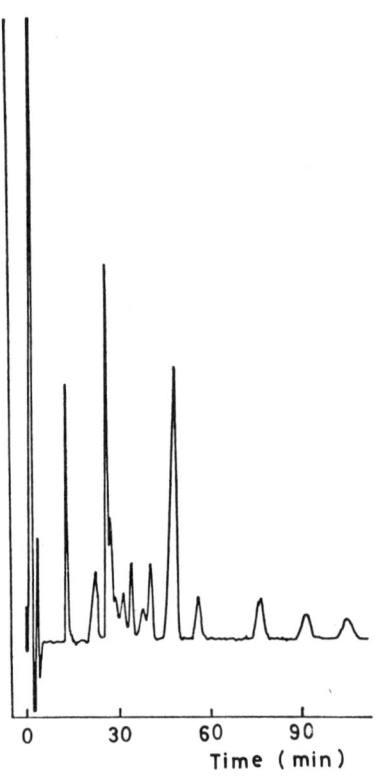

Detection of Amino Acids

Sample size: 2.5×10^{-7} mole (hydroxyproline: 5.0 and alanine: 22.5×10^{-7} mole); resin: Aminex A-4; column: 520 mm x 9 mm i.d., 40°C; eluant: 0.125M $CCIH_2COOH$-0.04M CH_3COOH- 0.1M CH_3COONa-0.1M NaCl; electrolyte: 0.01M CuSCN-4M KSCN; detection potential: 0.39 V vs. Ag-AgSCN(2M KSCN)

only the hydroxy group in estrone is oxidized. At higher potentials (+1.2V) both the keto and the hydroxy steroids are oxidized and both solutes give a similar response.

The detector cell designed by Takata and Muto (5) employed an electrode of carbon cloth or metal gauge lightly packed in the opening in a silicone rubber plate. This electrode was interposed between two other electrodes made of silver-silver iodide gauze also supported on carbon cloth. An example of a chromatogram of amino acids produced by this detector is shown in figure 4. It is seen that there is little dispersion of the solute bands and the sensitivities realized were very adequate for routine analytical work. Koen et al. (10) have designed a microcell polarographic detector employing a dropping mercury electrode and employed it for use in pesticide analysis, the authors claimed a sensitivity of 10^{-8} mol/l (ca. 10^{-8}-10^9 g/ml) and an example of a chromatogram they obtained for the separation of a mixture of p-nitroplenol, methyl parathion and parathion is shown in figure 5.

Figure 5

Detection of Pesticides by an Electrochemical Detector

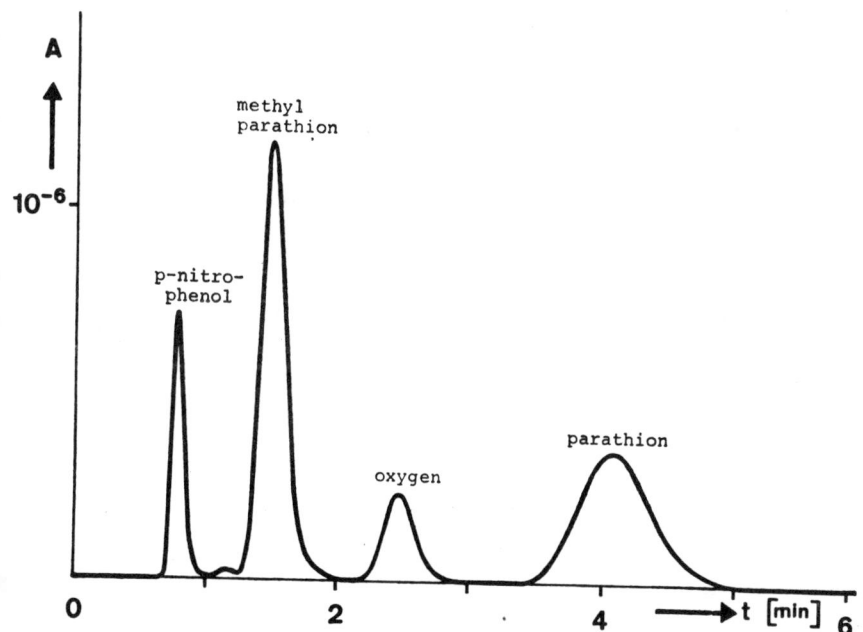

The polarographic detector has very specific areas of application. Sensitivities vary greatly with the solute detected and the detector can only be used for the detection of oxidizable or reducible solutes. In general the detector is relatively simple to use but at the time of writing this book and to the author's knowledge a commercial polarographic detector is not yet available. The problems associated with the use of polarography in the presence of oxygen can be eliminated when the technique is used for chromatographic detection purposes by employing the column to separate any oxygen in the sample from the solutes of interest. An example of this procedure is shown in figure 5. It is likely that the sensitivity of the polarographic detector could be increased with the use of more sophisticated equipment but to date this detector has not been readily accepted by chromatographers generally.

REFERENCES

1. W. Vermula, Rocznike Chem., 26 (1952) 281.
2. P. L. Joynes and R. J. Maggs, J. Chromatog. Sci., 8 (1970) 427.
3. A. MacDonald and P. D. Duke, J. Chromatog., 83 (1973) 331.
4. R. N. Adams, P. T. Kissinger, C. Refshange and R. Dreiling, Anal. Lett., 6 (1973) 465.
5. Y. Takata and G. Muto, Anal. Chem., 45 (1973) 1864.
6. R. N. Adams, Electrochemistry at Solid Electrodes , Dekker, New York, 1969.
7. J. F. Alder, B. Fleet and P. O. Kave, J. Electroanal. Chem., 30 (1971) 427.
8. B. Fleet and C. J. Little, J. Chromatog. Sci., 12 (1974) 747.
9. S. Yamada and H. J. Matsuda, Electroanal. Chem., 44 (1973) 189.
10. J. G. Koen, J.F.K. Huber, H. Poppe and G. den Boef, J. Chromatog.Sci., 8 (1970) 192.

CHAPTER 5

The Heat of Adsorption Detector

During the passage of a solute peak through a chromatograph column the solute is continuously adsorbed onto the stationary phase in the front portion of the peak and continuously desorbed in the rear portion of the peak. The adsorption-desorption effect results from the continuously increasing solute concentration in the front of the peak causing more and more solute to be driven onto the stationary phase to maintain solute equilibrium between the two phases and in the rear portion of the peak the converse occurs. As the solute concentration falls, so more solute is desorbed from the stationary phase to maintain solute equilibrium between the two phases. The absorption of a solute by the stationary phase is accompanied by the evolution of heat which results from the heat of solution of the solute in the stationary phase if it is a liquid and the heat of adsorption of the solute on the adsorbent if the stationary phase is a solid. Thus if the temperature is monitored at any point in the column, during the passing of a solute band, the temperature will rise above the column temperature as the front part of the peak passes the point of measurement and will fall below the column temperature as the rear part of the peak passes; subsequently the temperature will return to the original column temperature. This effect was employed by Claxton in 1959 to monitor liquid chromatography eluents.

The heat of adsorption detector, devised by Claxton (1), consists of a small plug of absorbent, usually silica gel, through which the chromatographic eluent passes subsequent to leaving the column. Embedded in the silica gel is either a thermocouple or a thermistor that continuously measures the temperature of the absorbent and mobile phase. When an eluted solute comes into contact with the silica gel, the heat evolved causes the temperature to rise. As the solute is subsequently desorbed from the silica gel, heat is adsorbed and the temperature falls. The output from the temperature measuring device thus records an increase in temperature and

140   Liquid Chromatography Detectors

then a decrease in temperature relative to its surroundings and an S-shaped curve results.

Figure 1

The Heat of Adsorption Detector Cell Developed by Zlatkis et al.

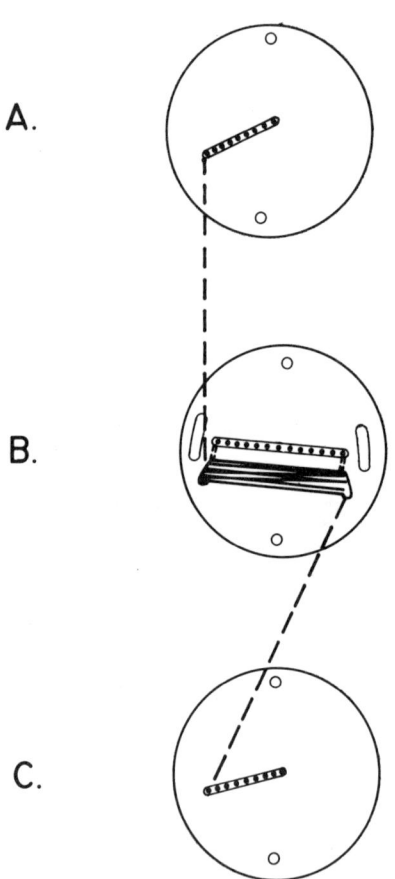

The teflon discs are 35 mm in diameter. A, Top teflon disc; B, Mounted thermopile (size of thermopile compartment: 3 mm x 3 mm x 20 mm; total length of thermopile: 17 mm); C, Bottom teflon disc with filter.

The heat of adsorption detector has been investigated by a number of workers (2,3,4), but, although commercially available, has not been extensively used as an LC detector. One reason for this is the curious and unpredictable shape of the temperature-time curve that results from the detection of the usual Gaussian or Poisson concentration profile of a peak eluted from the column. The shape of the curve changes with the operating conditions of the chromatograph, with the retention volume of the solute, and for closely eluted peaks, it produces a complex curve that is extremely difficult to interpret. A diagram of the heat of adsorption detector cell designed by Zlatkis et al.(4) is shown in Figure 1.

The detector is composed of multi-thermocouples (thermopile) with a suitable adsorbent embedded at one end which becomes the hot junction, with the opposite end serving as the cold junction. Iron and constantan, insulated with teflon, were used as the thermoelectric base wires. The wires were held parallel and wound around a plexiglass board 13 mm wide and 3 mm thick, so that the iron and constantan wires alternated. The position of the wires was fixed by covering them with a sodium silicate solution and allowing the solution to dry. After drying, the teflon was removed from the wire along the thickness of the board on each side with a razor blade and subsequently the wires were cut along the thickness of the board so that 1.5 mm of wire was exposed at each end. The exposed ends were welded together to form a group of thermocouples which comprised the thermopile. Two thermopiles were obtained from each operation and the number of thermo couples in each thermopile is equal to the number of winds made around the plexiglass board.

The thermopile was mounted in a teflon disc, 3 mm in thickness, and soldered to teflon-insulated leads which were connected to the input terminals of the amplifier. The mounted thermopile was then placed between two teflon discs, each 6 mm thick, and the three discs which form the sensing element were held together by two stainless steel pins. Each disc had a 1.00 mm groove extending from the center to a 1.00 mm hole through the disc so that the mobile phase entered the detector and flowed across the hot junction. A stainless steel filter was mounted in the teflon disc on the exit side of the thermopile to retain the adsorbent embedded in the exit end of the thermopile. The detector was situated between the two steel flanges containing cavities for the centering of the discs. The flanges were held together by means of 4 Allenhead screws, which also served to seal the system externally. Each flange contained two additional holes through which the teflon discs could be forced out of the cavities in which they rest. Each flange had a 1/8 in. O.D. x 0.027 in. I.D. stainless steel tube 1.0 in. in length soldered in the center which served to connect the detector to the column and to the eluate container. All connections were made with Swagelok fittings. The whole detector was thermostated and the output from the thermopile amplified and fed to a potentiometric recorder.

Figure 2

Chromatograms from the Heat of Adsorption Detector

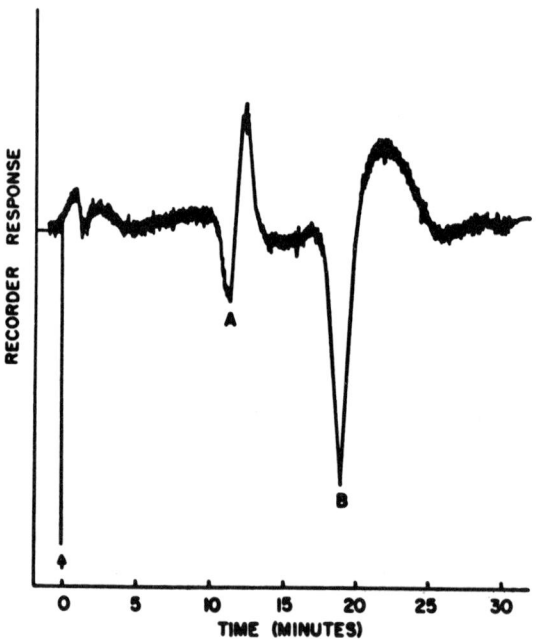

Separation of a hydroxy-acid from an ester using the heat of adsorption detector. Column: 4.5 mm I.D., x 6 ft. Stationary phase: Durapak (N-Octane Porasil C, 120/150 mesh). Mobile phase: Pentane. Temperature: 201°C. Pressure: 600 psi Flow rate: 1.5 ml/minute. A=Octyl Butyrate, 10 µg. B = 4-Hydroxyundecanoic acid, 10 µg.

A chromatograph obtained from this apparatus is shown in Figure 2.

The response of the detector has been examined by computer simulation by Smuts et al.(5) and an explicit equation describing the temperature change as a function of the concentration profile of the eluted peak and the thermal properties of the detector cell and column has been derived by Scott (6). The equation derived by Scott is as follows

$$\theta_v = \psi\, e^{-\phi v} \int_o^v e^{\phi v}\left[X_o \frac{e^{-v} v^n}{n!} - X_o \frac{e^{-v/C_\alpha}}{C_\alpha} \int_o^v -v/C_\alpha \frac{e^{-v} v^n}{n!}\, dv \right] dv$$

where θ_v is the temperature of the detecting cell, ψ is a constant, ϕ is a constant representing the heat loss factor, C_α is detector plate capacity-column plate capacity ratio, v is the flow of mobile phase through the detector in plate volumes of the attached column, and n is the efficiency of the attached column.

With the aid of the computer in the manner of Smuts et al. (5), the relative values of θ for v = 74 to 160 were calculated for a column having an efficiency of 100 theoretical plates, and for C_α taking values of 0.25, 0.5,

Figure 3

Temperature Curves and Integral Temperature Curves from the Heat of Adsorption Detector (Theoretical)

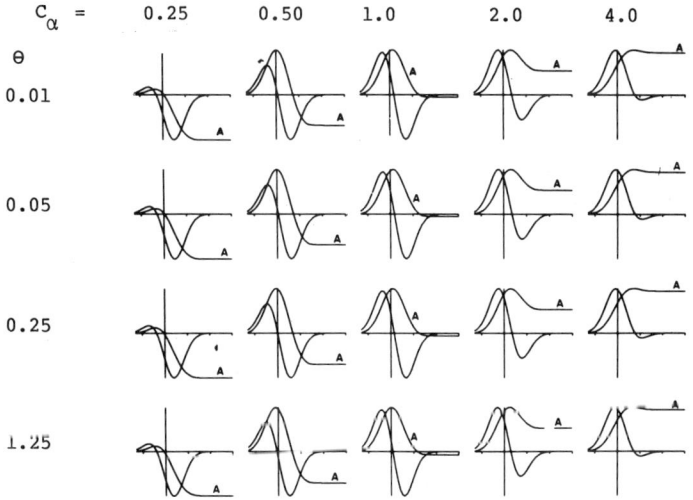

A   Integral Curves C_α Detector/Column Capacity Ratio
θ Detector Heat

1, 2 and 4, and for ∅ taking values of 0.01, 0.05, 0.25 and 1.25 respectively. The 20 curves shown in Figure 3 represent the shape of the Θ versus v_α curves and the integral Θ versus v_α curves for the different values of C_α and ∅ and are all normalized to the same peak height. The curves shown cover the practical range of heat loss factors for the detector cell, and demonstrate the effect of changes in detector capacity/plate capacity ratios that would result from different detection cell designs detecting a peak of constant width. The curves for different values of C_α would also represent the effect on peak shape of solutes of different retention, and thus different peak widths passing through a detecting cell of fixed dimension, but having an adsorbent different from the stationary phase of the column. Examination of Figure 3 shows that the major effect on peak shape is the ratio of cell capacity to column plate capacity C_α. It is seen that when the capacity of the detector cell is less then the plate capacity of the column ($C_\alpha < 1$), the negative part of the signal is predominant, whereas when the detector cell capacity exceeds that of the column ($C_\alpha > 1$), the positive part of the signal predominates. For this reason, the integral of the detector signal for $C_\alpha > 1$ rises to a peak but does not return to the baseline. For $C_\alpha < 1$ the integral curve first rises and then falls below the baseline and does not return. Only when $C_\alpha = 1$ does the detector signal simulate the differential form of the Gaussian curve, and its integral describe the true elution curve.

It follows that for the detector to be effective and useful, C_α must at all times be unity, and thus the detector must have the same plate capacity as the column for all solutes. This means that the detector must employ the same adsorbent, the same geometry, and be packed to have the same plate height as the column. It is obvious that to accomplish this, the column must be the detecting cell and the temperature sensing element be placed in the column packing itself.

If the end of the column is used as the detecting cell the temperature response is given by the following equation:

$$\Theta_n = \frac{\alpha}{\beta-1} \sum_{r=0}^{r=n} \left(\frac{\gamma-1}{\beta-1}\right)^r \frac{\partial X}{\partial v} (n-r)$$

where θ_n is the temperature of the n th plate of the column, α is a constant, γ is the heat loss factor of the cell due to convection, β is the heat loss factor of the cell due to conduction, n is the number of theoretical plates in the column, $X = \dfrac{X_o e^{-v} v^n}{n!}$, and v is the volume flow of mobile phase in column plate volumes.

Scott calculated a series of values for α, β and γ which are given below based on a pair of solutes eluted at k' values of 4 and 8 on a standard stainless steel column packed with silica gel employing heptane as the mobile phase.

k' = 4		k' = 8	
γ	1.50	γ	2.70
z	0.01	z	0.01
β	1.51	β	2.71
γ	1.50	γ	2.70
z	0.02	z	0.02
β	1.52	β	2.72
γ	1.50	γ	2.70
z	0.04	z	0.04
β	1.54	β	2.74
γ	1.50	γ	2.70
z	0.08	z	0.08
β	1.58	β	2.78
γ	1.50	γ	2.70
z	0.20	z	0.20
β	1.70	β	2.90

(Note $z = \beta - \gamma$)

Using the computer, the temperature and integral of the temperature curves were calculated employing the data given above. Values of β were chosen to cover a practical range of conditions used in liquid chromatography based on the approximate value for z of 0.02. The curves obtained are shown in Figure 4. All curves are normalized to the same peak height.

It is seen from Figure 4, that for a solute eluted at a k' value of 4, the value of z must be 0.08 if the differential form of the Gaussian curve is to be realized. Furthermore, for a solute eluted at a k' value of 8, the value of z must be increased to 0.2 if the integral of the temperature curve is to take the form of the true elution curve of the solute. These values for z are considerably greater than those calculated for the column. The value of z can be increased by improving the radial transfer of heat across the column. Z was only considered to be controlled by the thermal conductivity of the column contents, but the radial mass transfer process described by Knox (7) would also relate to the transfer of heat. However,

Figure 4

TEMPERATURE AND INTEGRAL TEMPERATURE CURVES FROM THE HEAT OF ADSORPTION DETECTOR WITH THE SENSOR SITUATED IN THE COLUMN PACKING (THEORETICAL)

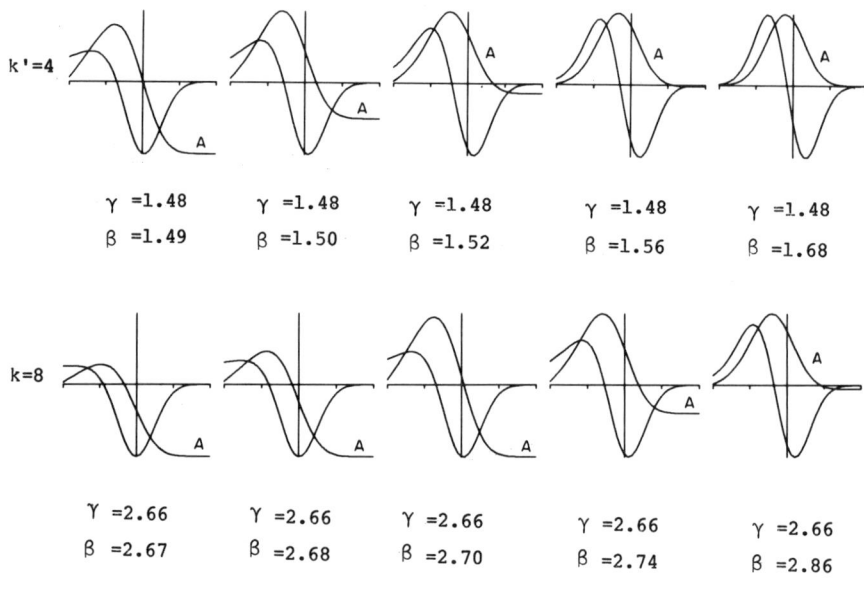

A = Integral Curves

even assuming the radial transfer of heat by the Knox principle to be equivalent to the heat transferred by conduction, this would only double the value of z. A practical value of 0.04 for z would only render the in-column detection method satisfactory for solutes eluted up to a k' value of 2 or 3.

From the theoretical treatment given the following can be concluded:
1) The output from a heat of adsorption detector will not provide the differential form of the elution curve of a solute unless the capacity of

the detecting cell is identical with the plate capacity of the column for all solutes. This condition can only be achieved in practice if the detector has the same cross-sectional geometry as the column, contains the same adsorbent, and is packed to give exactly the same efficiency (i.e., the same HETP). Thus, each detector would have to be constructed to suit each particular column.

2) If the column itself is used as the detector by inserting a sensing element into the packing, then the detecting cell will have the same capacity as the column for all solutes. However, under such circumstances there can be no heat exchanger between the last plate of the column and the detecting cell. In order to eliminate the effect of heat generated in the column prior to the detecting cell, which would be convected to the cell and distort its normal temperature profile, there must be adequate radial heat loss from the column. For normal packed columns it appears impossible, in practice, to achieve adequate radial heat loss that would provide a satisfactory signal from the sensing element. At best, an in-column heat of adsorption detector could only operate satisfactorily for solutes eluted at k' values below 2 or 3.

Bearing in mind other limitations of the detector, such as its inability to cope with gradient elution and temperature program methods of development, and that the adsorbent has to be changed from time to time, the practical difficulties of construction make the heat of the adsorption detector one of the least attractive solute property detectors for liquid chromatography. One possible area of application of this detector would be the microparticulate microbore liquid chromatography column (8). Such columns, constructed of stainless steel might have sufficient radial heat transfer if situated in a thermostating liquid that a thermocouple at the radial center and close to the column exit might give a true Gaussian differential curve for an eluted solute. Used in conjunction with an integrating amplifier an effective heat of adsorption detecting system providing a true representation of the elution curve might be feasible.

Although a commercial heat of adsorption detector was once available it is now no longer manufactured and thus no specifications are available for this detector.

REFERENCES

1. G. Claxton, J. Chromatogr., 2 (1959) 136.
2. A. J. Groszek, Nature, 182 (1958) 1152.
3. K. P. Hupe and E. J. Bayer, J. Gas Chromatogr., April (1967) 197.
4. J. L. Cashaw, R. Sigura and A. Zlatkis, J. Chromatogr. Sci., 8 (1970) 363.
5. T. W. Smuts, P. W. Richter and V. Pretorious, J. Chromatogr. Sci., 9 (1971) 457.
6. R. P. W. Scott, J. Chromatogr. Sci., 11 (1973) 349.
7. D. S. Knox and J. H. McLaren, Separation Techniques in Chemistry or Biochemistry , (Ed. R. A. Keller), M. Dekker Inc., New York.
8. R. P. W. Scott and P. Kucera, J. Chromatogr., 125 (1976) 251.

CHAPTER 6

The Spray Impact Detector

The production of electrical charges during the disruption of a liquid to produce an aerosol has been known for many years. As far back as 1892, Lenard (1) noted the ionization of air at the base of a waterfall where the water splashed against rocks. Christianson (2) investigated the electric charges produced when water was sprayed onto a solid plate and more recently Loeb (3) examined the effect in greater detail and termed the phenomenon "spray electrification". The charges are thought to be generated by the rapid break-up of a liquid surface either by a gas stream or by impact with a solid surface. Various mechanisms have been suggested to explain the production of electrical charge, but that generally accepted at this time is the one given by Mattison (4). Mattison suggests that any ions present in the liquid reside just below the surface and are not situated in the surface layer. If a new surface is formed by the sudden break-up of the original surface, the ions in the new surface are pulled inward by attraction from the bulk liquid, the force of which will depend on both the field associated with the ion and its mobility. The surface thus becomes rich in the slower, larger species of ion. For example, with water, the new surface would become richer in hydroxyl ions than hydrogen ions and thus would attain a net negative charge. In the author's opinion, there is still some uncertainty as to the exact mechanism of charge formation and for those readers interested in this subject they are recommended to refer to Mattison's original paper.

Mowery and Juvet (5) have employed the spray electrification effect in a novel form as a liquid chromatography detector, primarily for use with reverse phase liquid chromatography. A stream of the eluent from the column is allowed to strike a conducting target to form a spray and the potential of the electrode is monitored. A diagram of their detector is shown in figure 1. The target electrode is an electrically isolated conducting rod which is connected to a suitable electrometer that permits both the measure-

Figure 1

The Spray Impact Detector

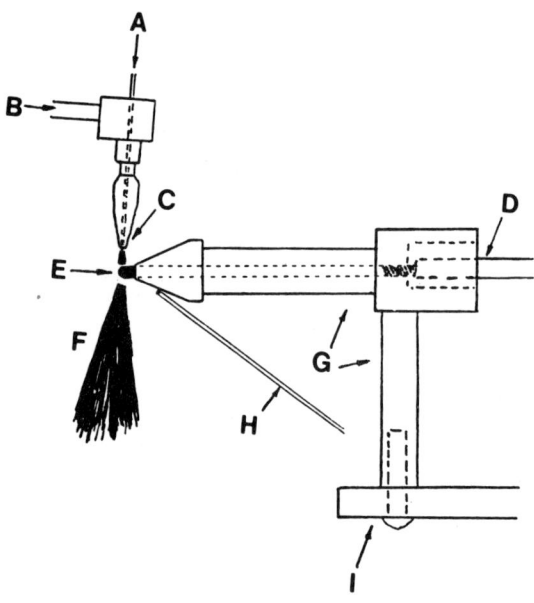

A, LC column outlet; B, air inlet to aspirator; C, stainless steel aspirator jet; D, electrometer coaxial cable with epoxi seal; E, glassy carbon or gold target electrode; F, spent spray from target; G, PTFE body and mounting arm; H, glass capillary to aspirator vacuum line for charged droplet removal; I, laminated plastic mounting.

ment of electrode potential and electrode current. Two types of electrodes were investigated, one composed of glassy carbon and the other of gold-plated platinum. A coaxial cable connected the electrode to the electrometer, via a soldered connection to the electrode which was supported on a high resistance laminated plastic mounting. On impact between the eluent and the target, the target becomes negatively charged and the resulting

Table I. Performance Characteristics of the Spray Impact Detector. Mobile phase, boiled distilled water.

Compound	Detection Limit, g/sec	Linear Dynamic Range log (base 10)	Target Material
n-Octanoic acid	7×10^{-11}	4.0	gold
n-Nonanoic acid	8×10^{-11}	4.0	gold
n-Decanoic acid	5×10^{-11}	4.0	gold
Trifluoroacetylacetone[a]	3×10^{-10}	4.0	gold
Trifluoroacetylacetone[b]	2×10^{-9}	3.0	gold
Ammonium 8-anilino-1-naphthalene sulfonate	9×10^{-10}	3.5	gold
o-Nitrophenol	9×10^{-10}	3.5	gold
o-Nitrophenol	2×10^{-10}	3.5	carbon
Sodium dodecylsulfate	5×10^{-10}	3.5	gold
Sodium dodecylsulfate	2×10^{-10}	3.5	carbon
Sodium dodecylsulfonate	2×10^{-10}	–	carbon
Sodium tridecylsulfate	5×10^{-10}	3.3	carbon
Ethylamine	2×10^{-9}	2.5[c]	carbon
n-Octanol	8×10^{-8}	–	carbon
n-Nonanol	8×10^{-8}	–	carbon
Ethyl butyrate	1×10^{-7}	> 2	carbon
2-Heptanone	2×10^{-7}	–	carbon

a Effluent flow rate, 3.1 ml/min.

b Effluent flow rate, 8.6 ml/min.

c Includes positive and negative areas (see text).

spray becomes positively charged. Small quantities of solute in the mobile phase changes this potential drastically and as an example, at a standing voltage of 2,000 volts, 1.5×10^{-7} g of nitrophenol present in the eluent was found to produce a voltage change of 400 volts. The detector system was situated in an earthed metal box through which filtered laboratory air was passed. This was necessary to reduce noise from air contaminants and reduce the effect of charged droplets that tended to accumulate in the neighborhood of the electrode. The performance characteristics of the system investigated by Mowery and Juvet are shown in Tables I, II, and III. In Table I, the detection limit and linear dynamic range of the detector for a series of organic compounds employing water as the mobile phase is given for the two different electrode types. Generally, it was found that the glassy carbon electrode gave a higher sensitivity and, furthermore, required no prior conditioning before use with aqueous mobile phases. It is seen from Table I that very high sensitivities were attainable with a linear dynamic range for most substances of 3 to 4 orders of magnitude. In Table II, the sensitivity

Table II. Performance Characteristics of the Spray Impact Detector for Some Inorganic Compounds. Mobile phase, boiled distilled water. Gold target electrode.

Compound	Detection Limit g/sec	Linear Dynamic Range log (base 10)
Lithium nitrate	1×10^{-9}	$2.7 + 1.7^a$
Potassium nitrate	9×10^{-10}	$2.8 + 1.5^a$
Lanthanum nitrate	4×10^{-10}	$3.0 + 1.5^a$
Thorium nitrate	5×10^{-10}	$3.0 + 1.5^a$

a Two linear regions of different slope with discontinuity occurring at zero current.

and linear dynamic range of the detector is given for a number of inorganic compounds again employing water as the solvent. Very high sensitivities are again realized and it is interesting to note the detector has two linear response ranges for each solute. In Table III, the sensitivity is given for a number of substances employing a range of different mobile phases. It appears that the solvent employed can have a profound effect on the detector sensitivity. Employing methyl ethyl ketone as an alternative to acetonitrile or an acetonitrile/ water mixture as a solvent reduces the detector sensitivity by two or three orders of magnitude. In a similar manner the sensitivity varies significantly with the type of solute. For example, sensitivities of 10^{-10} g/sec are realized for σ-nitrophenol contained in a

Table III. Performance Characteristics of the Spray Impact Detector Using Organic and Mixed Organic/Water Mobile Phases. Carbon target electrode.

Compound	Detection Limit g/sec	Peak Direction	Mobile Phase
Glycine	5×10^{-11}	+	acetonitrile[a]
Stearic acid	7×10^{-10}	+	acetonitrile[a]
Sodium tridecylsulfonate	3×10^{-10}	+	12.5% acetonitrile/water
Sodium tridecylsulfonate	8×10^{-11}	+	25% acetonitrile/water
o-Nitrophenol	2×10^{-7}	+	acetonitrile[b]
o-Nitrophenol	1×10^{-10}	+	15% acetonitrile/water
Sodium cholate	3×10^{-11}	+	15% acetonitrile/water
p-Ethylphenol	1×10^{-7}	+	15% acetonitrile/water
n-Heptanol	3×10^{-7}	+	15% acetonitrile/water
Thorium nitrate	4×10^{-8}	−	1 mg/ml sodium dodecylsulfate/water
Water	3×10^{-6}	+	methyl ethyl ketone
Dimethylformamide	1×10^{-6}	+	methyl ethyl ketone
Ethyl acetate	4×10^{-5}	−	methyl ethyl ketone
n-Heptane	7×10^{-6}	−	methyl ethyl ketone
Toluene	3×10^{-5}	−	methyl ethyl ketone
n-Octylaldehyde	1×10^{-6}	−	methyl ethyl ketone
n-Octanol	2×10^{-8}	−	methyl ethyl ketone
o-Nitrophenol	2×10^{-7}	+	methyl ethyl ketone
Tetrachloroethylene	7×10^{-6}	−	methyl ethyl ketone
2-Heptanone	1×10^{-5}	−	methyl ethyl ketone

a Target first conditioned by wetting with water.
b No water conditioning of target electrode.

Figure 2

Chromatograms from the Spray Impact Detector

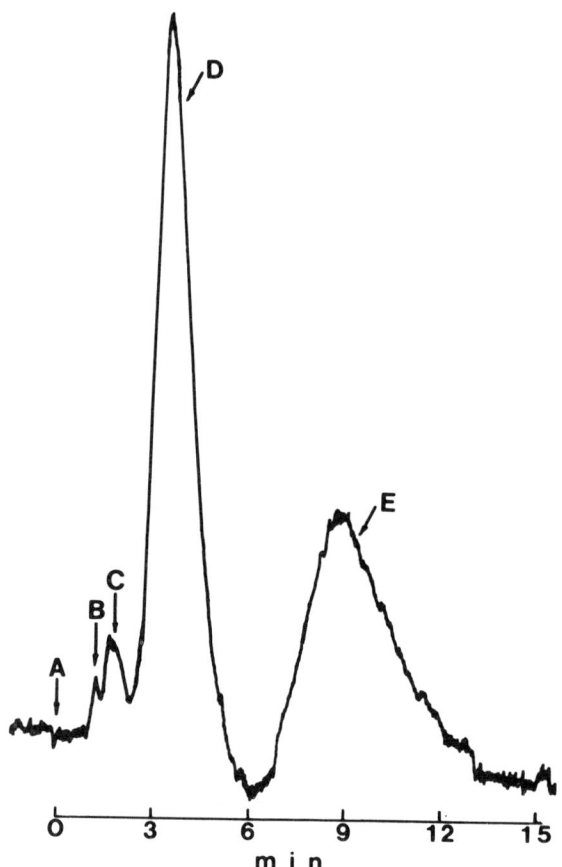

Separation of the n-C_8, C_9, and C_{10} fatty acids. Column, 60 x 0.6 cm Bondapak C_{18}/Corasil; target, gold; mobile phase, boiled distilled water at 5.9 ml/min flow rate; air flow, 4550 ml/min.
A, injection point; B, inorganic and organic impurities; C, 7.2 ng octanoic acid; D, 120 ng nonanoic acid; E, 80 ng decanoic acid.

15% v/v solution of acetonitrile in water, but in the same solvent the sensitivity of the detector to p-ethylphenol is only 10^{-7} g/sec. A chromatogram obtained with this detector for a mixture $n-C_8$, $n-C_9$ and $n-C_{10}$ fatty acids is shown in figure 2. The high sensitivity of the detector to these type of substances is clearly indicated.

This detector is obviously in its early stages of development and no commercial model is available at present. Some further work on it will be necessary before it would become a routine instrument for liquid chromatography purposes. However, it does appear to hold exciting prospects in the future, particularly for use in ion exchange chromatography. The extra column dispersion that is produced from this type of detector is, at this time, uncertain but it would appear that it should be possible to reduce or maintain any dispersion at an acceptable low level.

REFERENCES

1. P. Lenard, Ann. Phys. (Leipzig) Ser. 3, 46 (1892) 584.
2. C. Christianson, Ann. Phys. (Leipzig) Ser. 4, 40 (1913) 107.
3. L. B. Loeb, Static Electrifications, Springer-Verlag, Berlin, 1958.
4. M. J. Mattison, J. Colloid Interface Sci., 37 (1971) 879.
5. R. A. Mowery and R. S. Juvet, Jr., J. Chromatog. Sci., 12 (1974) 687.

CHAPTER 7

The Radioactivity Detector

The use of radioactive tracers in the study of reaction mechanisms has been steadily increasing over the last two decades. Radioactive tracers have been used to elucidate the mechanism of complex laboratory reactions, in photosynthetic chemistry and, in particular, to follow metabolic pathways of substances, synthetic and natural, both in plants and animals. In fact, the first reported use of an in-line radioactivity detector fitted directly to a gas chromatograph was in a paper by James and Piper (1) who developed the detector to study the synthesis of lipids, glycerides and fatty acids in plant tissue. James and his co-worker designed a small volume radioactivity counter that was placed in line with the column eluent but subsequent to an oxidizing furnace. The solute from the column was combusted to CO_2 and water and the tritiated water and radioactive carbon dioxide from the carbon fourteen tracer reduced in a second furnace containing iron and was then counted during passage through a Geiger-Muller tube. By integrating the output from the Geiger counter a measure of the total activity of the peak could be obtained. In liquid chromatography, however, as the mobile phase is a liquid, the same procedure cannot be used, and an alternative method of counting has to be resorted to.

In 1961, Schram and Lombaert (2) designed a flow through cell for the continuous radioactive monitoring of liquid streams by scintillation counting utilizing anthracene powder as the scintillation agent. The counting cell was made from a 60 cm length of polyethylene tubing 2.2 mm I.D. and 3.2 mm O.D., which was filled with anthracene powder by a slurry packing technique employing a suspending solvent composed of 30% v/v of water in ethanol. The packed tube was then coiled into a spiral, which was placed in a flat Lucite vial containing silicone oil. The size of the anthracene particles affected both the pressure drop across the tube and its counting efficiency. The authors compromised on the particle size, using particle diameters of 300 micron for carbon fourteen counting and 150 micron for tritium counting. The cell volume was about 1 ml which, although band dispersion was significantly reduced due to the presence of the anthracene packing, would be far too great for modern high efficiency microparticulate columns. The detector tube was situated in a light sealed box and counting was achieved using one or more refrigerated photomultipliers. The counting efficiency obtained was about 2% for tritium and about 55% for carbon-14. The background gave about 60 counts/min. At a signal-to-noise ratio of two 1.5 nC/ml of tritium and 0.05 nC/ml of carbon-14 could be detected.

In 1964, Sjöberg and Agren (3) described a dual in-line adsorption and radioactivity detecting system. This type of system is of great practical value as the output from the adsorption detector discloses the position and relative proportion of all the solutes present in the mixture whereas the radioactivity detector monitors only those solutes that are radioactive. The radioactivity detector cell of Sjöberg and Agren consisted of a commercially available plastic scintillation tube in the form of a spiral, 1 3/4 in. in diameter composed of a single tube 2 ft. long, 0.7 mm I.D. and 1.5 mm O.D. having a volume of 0.3 ml. Such a cell would cause very serious band dispersion if employed with high efficiency columns. In the original paper by Sjöberg and Agren, an excellent example is given of the use of the two simultaneous detecting systems in the separation of some acid soluble nucleotides extracted from diploid tumor ascilis cells 30 min after incubation with radioactive inorganic phosphate.

The radioactive detecting systems devised by Schram (2) and Sjöberg (3) were reasonably efficient for counting carbon-14 and isotopes having β-particle emission of higher energy but their efficiency for counting tritium was relatively poor. Furthermore, some compounds could be adsorbed by anthracene, thus causing a build-up of background noise and loss of such compounds. The cells could also only be used with solvents which do not affect the material of the cell, anthracene or the scintillator. Scharpenseel and Menke (4) attempted to improve the efficiency of the tritium count by employing a toluene based liquid scintillation system. The column eluent or a portion thereof is mixed with the reagent and then passed to the scintillation cell. Counting efficiency for tritium was increased to 2 to 5%, but the system was extremely sensitive to the presence of salts even when the effluent to scintillator ratios were very low. Hunt (5) attempted to improve this system by replacing the toluene based scintillator with a solution of naphthalene 2,5-diphenyloxazole and (1,4-bis-2(4-methyl-5-

phenyloxazole)benzene) in carefully purified dioxane. The cell Hunt used was a coiled tube similar to that of Schram (2) so that band dispersion again would be a serious problem if high column efficiencies were required. Hunt, however, achieved a counting efficiency for tritium of about 14% and about 70% for carbon fourteen.

Figure 1

Flow Cell for Heterogeneous Radioactivity Counting

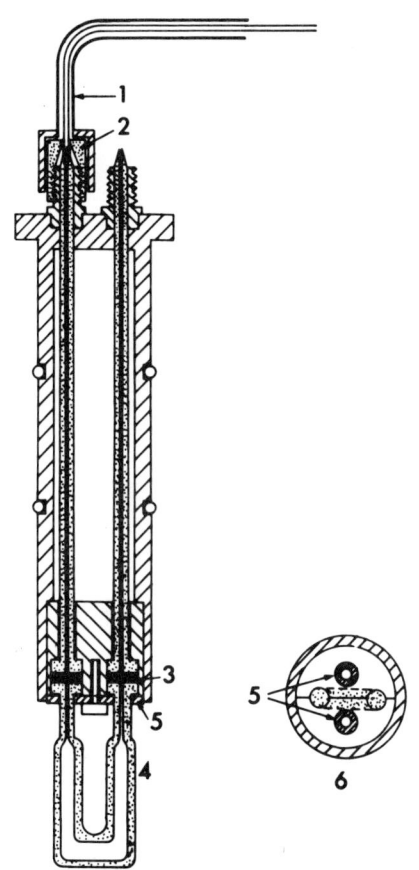

1 - Metal cap and bent sleeve for light-proofing; 2 - Swagelok fitting;
3 - Viton O-rings; 4 - interchangeable borosilicate glass cell;
5 - semi-circular collar disks; 6 - bottom view of cell and probe.

Schutte (6) examined the methods of Hunt and Sjöberg and developed a dual system of adsorption and radioactivity detection. Schutte also introduced a heterogeneous scintillation system employing cerium activated lithium glass beads as the scintillation agent. A diagram of the cell employed by Schutte is shown in figure 1. The U tube at the base is packed with the scintillator beads and is connected directly in line with the column eluent. The counting efficiency of this system is relatively poor, 0.2% for tritium and 17% for carbon-14. The advantage of this method is that it lends itself to the design of cell and connecting tubes that would provide minimum band dispersion due to the fact that the counting cell is packed and therefore could possibly be used with high efficiency microparticulate columns. Another form of radioactivity detection is that based on a solute transport system (7) but this will be discussed under solute transport detectors.

REFERENCES

1. A. T. James and E. A. Piper, J. Chromatogr., 5 (1961) 265.
2. E. Schram and R. Lombaert, Anal. Biochem., 3 (1962) 68.
3. C. J. Sjöberg and G. Agren, Anal. Chem., 36 (1964) 1017.
4. N. W. Scharpenseel and K. M. Menke, Tritium Phys. Biol. Sci., Proc. Symp., 1961 (Publ. 1962) 281.
5. J. A. Hunt, Anal. Biochem., 23 (1968) 289.
6. F. Schutte, J. Chromatogr., 72 (1972) 303.
7. N. Dugger, American Patent, filed November 29th, 1974, No. 528343.

CHAPTER 8

The Electron Capture Detector

The electron capture detector only gives a significant response to electron capturing substances and, therefore, even when used as a gas chromatographic detector it will only detect certain classes of compounds. A significant proportion of the solvents used in liquid chromatography do not give a response with the electron capture detector and, therefore, such

Figure 1

The Electron Capture Detector Designed by Nota and Palombari

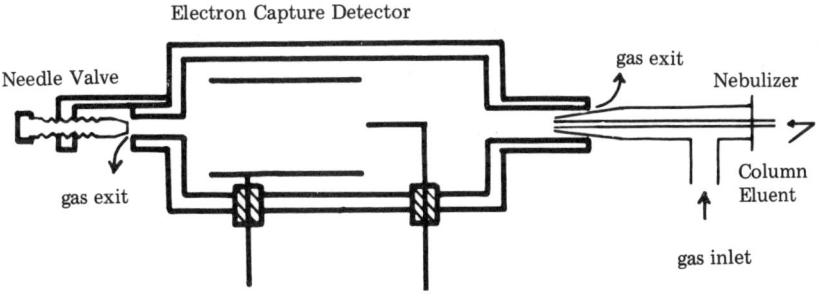

solvents can be volatilized along with any solute that is present, passed through the detector and it will respond to the solute only, providing it has electron capturing properties. This principle was developed by Nota and Palombari (1) to give an effective liquid chromatography detector. Such a detector, however, is very selective but as a number of important classes of compounds (e.g., pesticides, carcinogens, etc.) are electron capturing in nature, this detector can be a very valuable aid in pollution studies and food analysis.

A diagram of the apparatus developed by Nota and Palombari is shown in figure 1. The eluent from the column passes directly into an atomizer, the outlet tube from the column terminating at the nozzle of the atomizer. A portion of the atomized eluent passes directly into an electron capture detector and then out to waste. The electron capture detector is operated

Figure 2

The Pye Unicam Liquid Chromatography Electron Capture Detector

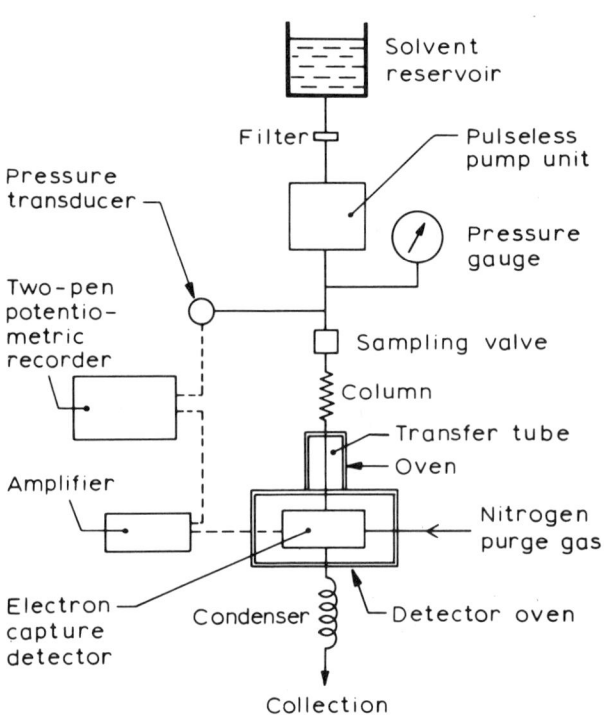

under the same conditions as those that would be used if it were employed as
a gas chromatographic detector. The work of Nota and Palombari established
the system as a viable liquid chromatographic detector and showed that the
solvents benzene, hexane, cyclohexane, pyridine, methanol, ethanol, diethyl
ether and acetone could all be employed as solvents without significantly
affecting the standing current of the detector. The disadvantage of this
system was that by employing an atomizer, it was necessary to use relatively
large volumes of nitrogen and further the interface between column and
detector introduced significant band dispersion.

Willmott and Dolphin (2) developed an improved form of the detector and
a diagram of their detecting system is shown in figure 2. It is seen that
the eluent is vaporized directly into the detector in an atomized form by
means of a heated transfer tube situated in an oven.

On leaving the column, the eluent passes into a stainless steel transfer
tube (1000 mm x 1.59 mm o.d. x 0.25 mm i.d.), enclosed in the oven, the
temperature of which (300°C) is such that the liquid is completely vaporized. The increase in volume involved in this transition, forces the vapor
into a ^{63}Ni electron capture detector, which is maintained at 300°C in the
same oven as the transfer tube. A purge of 30 ml/min of nitrogen sweeps the
vapor through the detector into a coil of stainless steel tubing (2000 mm x
2.3 mm i.d.) from which it is collected as a liquid.

The signal from the electron capture detector is amplified using either
a pulse mode amplifier or a constant current amplifier. The latter can
operate with 60V pulses having a nominal width of 1 μsec over a frequency
range 0-130 kHz. The standing current could be varied between zero and 5 x
10^{-9} A.

The detector described by Willmott and Dolphin is now manufactured
commercially by Pye Unicam, Ltd. The data available for this detector is
shown in Table I. It is seen from Table I that the sensitivity to electron
capturing substances is extremely high (1.2 x 10^{-10} g/ml) but the linear
dynamic range is only about 100. Further, the dynamic range of the detector
is only about one order of magnitude greater than its linear dynamic range.
However, the high sensitivity of the detector makes it an extremely valuable
aid in the trace analysis of pesticides and other similar substances. A
chromatogram of pesticides from the electron capture detector by Willmott
and Dolphin is shown in figure 3. The chromatograms shown are obtained from
the electron capture detector and a UV detector placed in series with the
column eluent. It is seen that 3.2 x 10^{-8} g of lindane is clearly detected
by the electron capture detector whereas the UV detector exhibits no

Table I

Specifications for the Pye Unicam Electron Capture Detector

Dynamic Range	ca. 1000
Response Index	Not given
Dynamic Range	ca. 100
Detector Response	Not given
Noise Level	4×10^{-12} A
Detector Sensitivity	1.2×10^{-10} g/ml
Shape	Not applicable
Dimensions	Not applicable
Connecting Tube	Length 100 cm
	Diameter 0.25 mm
	Volume 49 ul
Time Constant	Not given

response at all. This combination again demonstrates the advantages resulting from the use of two detectors to simultaneously monitor the column eluent. In using two or more detectors in this way, however, requires that the interconnecting tube does not contribute significantly to band dispersion.

Figure 3

The Comparative Performances of the Ultraviolet (UV) and Electron Capture (ECD) Detectors

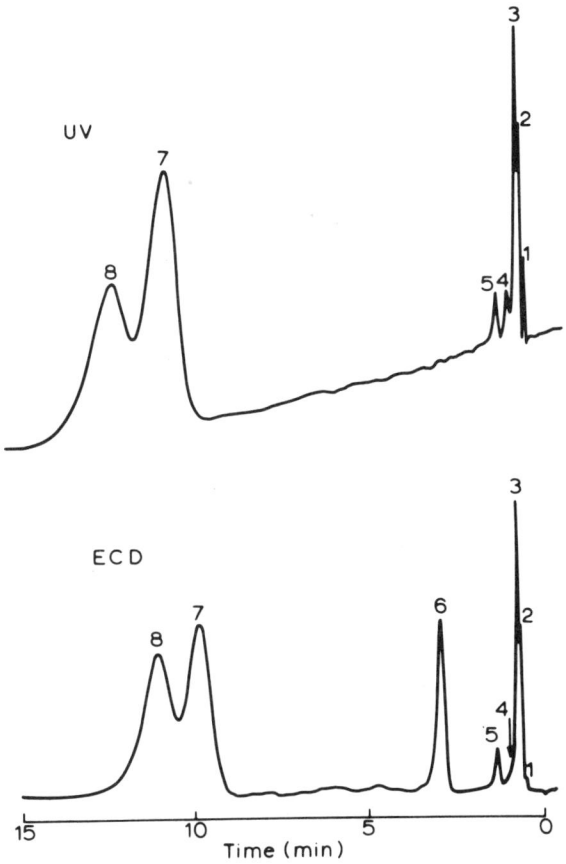

1. heptachlor impurity, 2. aldrin, 3. heptachlor, 4. heptachlor impurity, 5. heptachlor impurity, 6. lindane, 7. endrin, 8. dieldrin.
UV Chromatogram: detector range 0.05 a.u., wavelength 230 nm. Sample composition: aldrin 1.4×10^{-7} g, heptachlor 2.9×10^{-7} g, lindane 4.0×10^{-7} g, endrin 13.5×10^{-7} g, dieldrin 12.7×10^{-7} g.
ECD Chromatogram: Attenuation X512, standing current 0.73×10^{-9} A, purge 30 ml/min N_2.
Sample composition: aldrin 1.1×10^{-8} g, heptachlor 2.3×10^{-8} g, lindane 3.2×10^{-8} g, endrin 10.8×10^{-8} g, dieldrin 10.1×10^{-8} g.

REFERENCES

1. G. Nota and R. Palombari, J. Chromatogr., 62 (1971) 153.
2. F. W. Willmott and R. J. Dolphin, J. Chromatogr. Sci., 12 (1974) 695.

CHAPTER 9

Transport Detectors

In most liquid chromatography detectors, the sensing system monitors some property of the solute that is not shared by the solvent or that the solvent has to markedly less extent. It follows that such detecting systems are, to some extent, selective in their detecting capabilities and further restricts the choice of solvents to those that do not possess the property being measured. Thus, the choice of mobile phase is limited, and this can be particularly disadvantageous when employing gradient elution development. The transport type of detector was developed to overcome these limitations.

The transport detectors consist of a carrier that can be a metal chain, wire or disc that continuously passes through the column eluent taking a sample with it as a thin film of mobile phase adhering to its surface. The mobile phase is then removed, usually by evaporation, leaving any solute contained in the mobile phase as a coating on the carrier. The carrier is then examined by a suitable detecting procedure to monitor the solute alone. If a flame ionization detector is employed to monitor the solute, it will detect any substance containing carbon and permit the use of any solvent providing it is reasonably volatile. The only restriction of this system is that the solute must be involatile, or it would be lost during the evaporation of the mobile phase. The system appears ideal, but there are some disadvantages, the main one being a relatively low sensitivity (ca.10^{-6} g/ml), but it appears with improved design, this sensitivity could be significantly improved.

The moving wire detecting system was originally developed by James, Ravenhill and Scott (1), who eventually patented the system in 1965. The transport detector was subsequently made commercially available and a diagram of the Pye Unicam moving wire transport detector is shown in figure 1. Wire from a spool passes through a cleaning furnace maintained at 750°C then round a pulley and through the effluent stream of mobile phase from the

Figure 1

The Wire Transport Detector (By Pyrolysis)

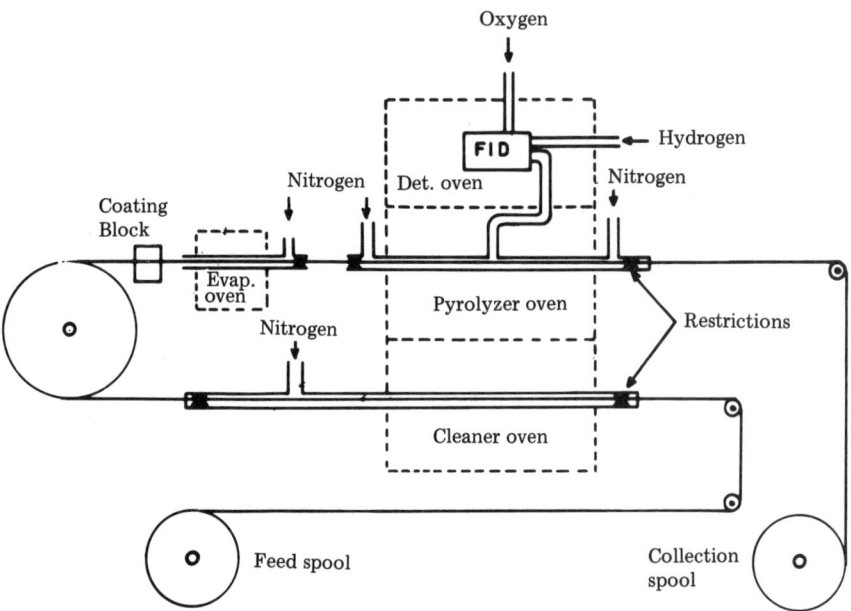

chromatographic column. After being coated with mobile phase, the wire passes into an evaporator oven at a temperature of about 105°C and then into a pyrolysis furnace. A nitrogen stream sweeps the solvent vapor out of the evaporator and on entering the pyrolysis furnace, two nitrogen streams from either end of the pyrolysis furnace sweep the pyrolysis products from the solute on the wire directly into a flame ionization detector. The output from the detector is fed to a suitable amplifier and then to a recorder. The sensitivity of the detector was about 5×10^{-6} g/ml but varied with the nature of the substances detected. Furthermore, the linear dynamic range was less than two orders of magnitude. Although establishing the viability of the detecting system, the performance left a lot to be desired as an effective liquid chromatography detector.

About the same time as the development of the wire transport detector,

Haahti and Nikkari (2) described a similar device, but more simple in design, employing a chain loop in place of wire. A diagram of their apparatus is shown in Figure 2. A gold chain, driven by a synchronous motor passes over a coating block where the chain is wetted with the column eluent

Figure 2

The Chain Transport Detector

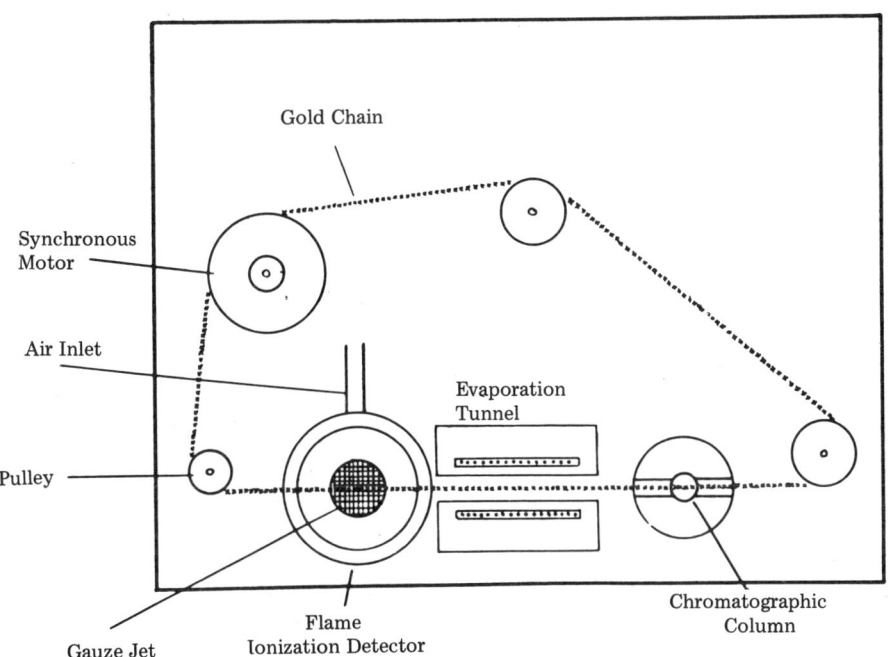

and then into an evaporator. The mobile phase is removed by evaporation, and the chain then passes directly through the flame of a flame ionization detector.

During combustion of the solute in the hydrogen flame, ions are produced in the normal way, and the resulting ionic current amplified and fed to a suitable recorder. Due to the occlusion of local, high concentrations of solute between the links of the chain, the detector was extremely noisy and, thus, provided relatively low sensitivity. A chromatogram obtained from the moving chain detector is shown in figure 3. The noise spikes on the peak are clearly seen, which besides affecting the overall sensitivity of the detector, also rendered quantitative analysis difficult. The advantage of

Figure 3

Chromatogram of Mineral Oil and a Surfactant Using the Chain Detector

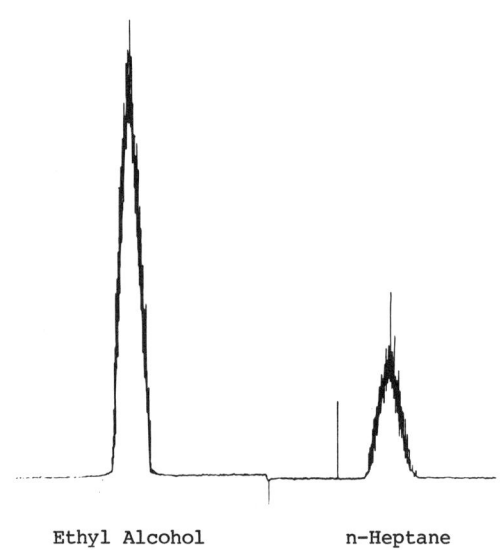

Ethyl Alcohol   n-Heptane

Column 2 x 300 mm packed with silica gel, flow rate 0.7 ml/min

the moving chain detector was, in its simplicity, relative to the moving wire detector.

Many workers in the field examined and developed the transport detectors and, in 1966, Karmen (3) introduced an aspirating system to draw the pyrolysis products into the hydrogen flame detector. In 1970, Scott and Lawrence (4) developed the system of Karmen further and introduced a modified form of detector in conjunction with the FID. The full sensitivity of the moving wire detector employing a pyrolysis system is only realized for certain compounds, for example, high boiling hydrocarbons, such as squalane or long chain fatty acids, such as stearic acid. For highly oxygenated compounds, such as carbohydrates, polyglycols, etc., the sensitivity of the detector may be reduced by as much as an order. It is evident that, if the senstivity of the wire detector could be either increased or maintained for oxygenated substances, it would be a far more useful detector in liquid chromatography.

The sensitivity of the wire detector, besides being dependent on the noise level of the system, also depends on the quantity of volatile pyrolysis products produced from the solute. Excluding synthetic polymers, which often quantitatively produce monomers on pyrolysis, many compounds only yield a few percent of their mass as volatile combustible pyrolysis products. Thus, the flame ionization detector (FID) may only, in effect, detect a few percent of the solute carried into the pyrolyzer by the moving wire. If, however, instead of pyrolyzing the solutes, they were completely

Figure 4

The Wire Transport Detector (By Methane Conversion)

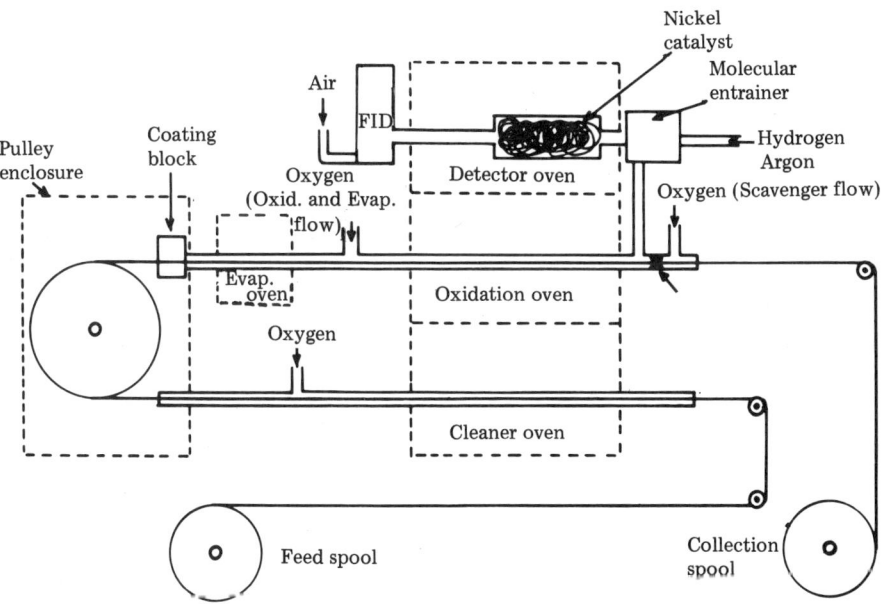

combusted in an oxygen or air stream, then all the carbon in the solute would be converted to carbon dioxide. Further, if the carbon dioxide was then reduced to methane by mixing with excess hydrogen and passing over a nickel catalyst, all of the carbon in the solute would be detected as methane by the FID. Such a system could increase the sensitivity of the detector to compounds that give low yields of volatiles on pyrolysis. Furthermore, potentially this detecting system should have a wide linear dynamic range and a predictable response.

A diagram of the moving wire detector modified in this way and developed by Scott and Lawrence is shown in figure 4. The FID detector was modified by enlarging the hydrogen lines in the detector bodywork to an internal diameter of 5/32 in. and the jet diameter to 0.070 in. This was necessary to reduce the flow impedance of the detector system and permit the satisfactory operation of an aspirator. The normal pyrolysis nitrogen inlet was blanked off and the detector connected to a 2 in. length of 1/2 in. diameter thin-walled stainless steel tube, which was connected to the aspirator by means of a length of 1/4 in. tube. The 1/2 in. tube was closed with a loose wad of quartz wool and filled with about 2 g of nickel catalyst. The nickel catalyst was prepared by adsorbing a saturated solution of nickel nitrate on 20/40 BS mesh brick dust, decomposing the nitrate to the oxide by heating to 500°C for 3 hours and reducing the oxide to metallic nickel in a stream of hydrogen at 250°C. The aspirator consisted of a jet and a Venturi and is placed in line with the hydrogen flow to the detector. The passage of hydrogen from the jet to the Venturi resulted in a pressure drop around the Venturi, and thus, allowed other gases to be sucked continuously into the hydrogen stream.

The reduced pressure side of the aspirator was connected to the side limb of the oxidation tube by means of a silicon rubber sleeve. It is seen from figure 4 that the two-tube system in the normal wire detector has been replaced by a single tube. The oxygen or air is fed in at the center of this tube, providing both the evaporator flow and the oxidation flow. The oxidation and cleaning tubes were constructed from quartz tube 4 mm O.D., 2.5 mm I.D., the restriction having an I.D. of 0.018 in. The oxidation tube contains only one restriction at the end of the system. Oxygen or air is passed into the tube after the restriction to prevent atmospheric air containing carbon dioxide entering the detecting system. The tube extended from the coating block to the final pulley and, thus, none of the wire was exposed to the air after having been coated. The cleaner oven tube was also arranged so that it extended from the first pulley to the pulley prior to the coating block. This again minimized contact of the clean wire with the atmosphere and, thus, reduced the noise. Oxygen or air was used for the cleaner flow to oxidize the material adsorbed on the wire and clean it prior to coating.

The linear dynamic range of the system was shown to be about four orders of magnitude as indicated by the curve in figure 5. The response index

Figure 5

Graph of Log I (Ionization current) Against Log (CO_2 concentration) for the Modified Moving Wire Transport Detector

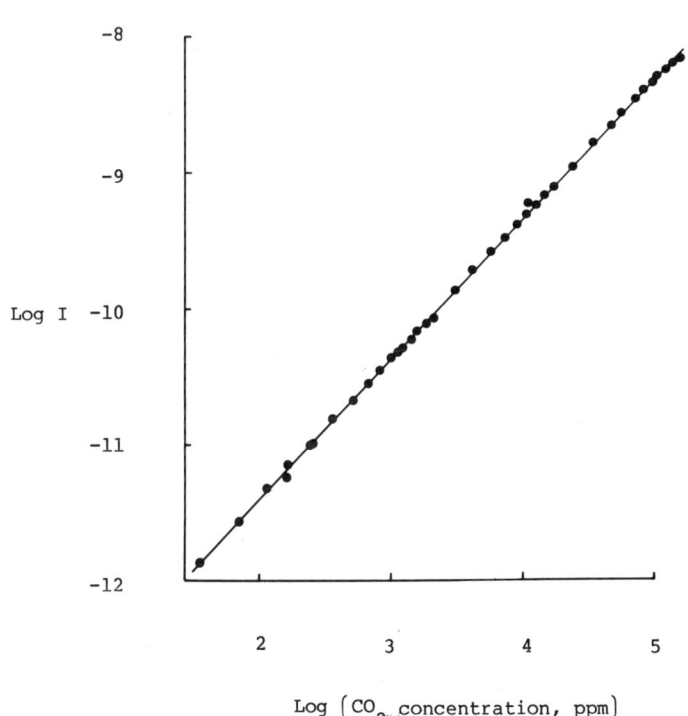

Table I

Specifications of the Pye Unicam Moving Wire Detector

Dynamic Range	10^5
Response Index	0.96 - 1.04
Linear Dynamic Range	10^4
Detector Response	Proportional to % carbon in solute
Detector Sensitivity	2×10^{-6} mm/ml of squalane
Cell Shape	Not applicable
Cell Dimensions	Not applicable
Connecting Tube	Length: 5 mm
	Diameter: 0.25 mm
	Volume: 0.25 µl
Amplifier Time Constant	Not given

determined for a series of compounds of different chemical types was found to be between 0.96 and 1.04. This instrument is now made commercially by Pye Unicam, Ltd., and the specification of the commercial model is given in Table I. It was found by Scott and Lawrence that the response of the detector was proportional to the carbon content of the solute and, thus, if the percentage of carbon in the solute was known prior calibration was not necessary. The response of the detector to carbon content, however, was only tested with a limited number of compounds and so this relationship should only be assumed with caution. A chromatogram of blood lipids obtained from the Pye Unicam Instrument employing incremental gradient elution development is shown in figure 6.

Van Dijk (5) developed a spray procedure for coating the wire in an attempt to improve the sensitivity of the detector. The column effluent passed directly to an atomizer, the nozzle of which was situated directly above the wire and one to two millimeters from it. The effect of spray coating the wire was firstly, to concentrate the solute in the mobile phase,

Figure 6

Chromatogram of Blood Lipids Using the Modified
Moving Wire Transport Detector

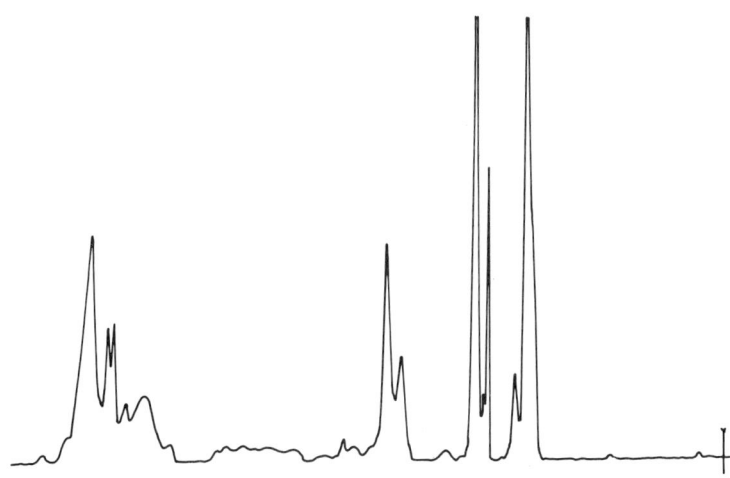

Incremental Gradient Elution Development

due to partial evaporation of the solute during atomization and, secondly, to increase the load on the wire by the formation of droplets. A linear response was obtained from this system and was shown to have a linear dynamic range of about 3×10^3. The author also claimed a sensitivity increase of about 50 over the conventional pyrolysis moving wire detector. It is difficult to determine from the publication the exact sensitivity that was obtained but from the calibration curves that were given, the sensitivity appeared to be about 3×10^{-6} g/ml. Stolyhwo, Privett and Erdahl (6) attempted to improve the sensitivity of the detector by using metal spirals wound on wire and stranded wire to increase the surface area of the carrier and, thus, increase the proportion of the column eluent taken into the detector. The authors claim a sensitivity limit for triolein of 100 nanograms. Again, it was not clear from the publication the exact volume of mobile phase in which this mass of solute was contained. If the 100 nanogram of triolein was eluted in a peak 1 ml wide at the base, the concentration at the peak maximum would be twice the average concentration i.e. 2×10^{-7} g/ml, which, for a transport detector, would be a great improvement on sensitivity. If, however, the same mass was eluted as an early peak in the chromatogram with a band width of only 50 µl, then the sensitivity would only be 4×10^{-6} g/ml, which would be no better than the previously developed transport systems. This uncertainty emphasizes the importance of specifying sensitivity in terms of minimum detectable concentration, which allows the direct comparison of the sensitivity of one detector with another.

Pretorius and Van Rensburg (7) attempted to increase the quantity of column eluent taken on to the carrier by coating the wire with sodium silicate, kaolin and copper kaolin. Sensitivity was again not quoted in terms of minimum detectable concentration, so the precise change in sensitivity that resulted from the coated carrier is not clear. From this publication, however, it would appear that a significant improvement in sensitivity was realized. The introduction of the wire coating procedure, however, further complicates an already complex instrument. It would seem that this approach might lead to serious instrumental problems arising from the dust produced by the disruption of the coating from abrasion as the wire passed round the pulleys and on rewinding the wire on the spool.

Šlais and Krejčí (8) replaced the normal flame ionization detector with the alkali flame ionization detector to selectively detect chlorinated compounds. These workers used a combustion technique as opposed to pyrolysis, mixed the products of combustions with hydrogen, and then passed the mixture directly to the alkali flame ionization detector. At a column flow rate of 0.37 ml/min, the sensitivity of the detector was stated to be 3×10^{-7} g/sec, which is equivalent to a sensitivity of about 1.6×10^{-6} g/ml. The moving wire detector has also been modified to provide radioactivity detection by Dugger (9) for monitoring tritium or carbon-14 labelled compounds. To detect carbon-14 compounds, the solute on the wire was oxidized to carbon dioxide, and the radioactive gas passed to a Geiger-Muller tube. To detect tritium, the tritiated water produced during oxidation of the solute was then passed over heated iron to reduce it to hydrogen, which was then also passed through a Geiger-Muller tube. Specificiations and performance characteristics of the apparatus were not given.

The wire transport system has many attractive characteristics as a liquid chromatography detector, but for general use, its sensitivity needs to be increased by at least an order of magnitude. Further, the overall

Figure 7

The Dubský Disc Detector

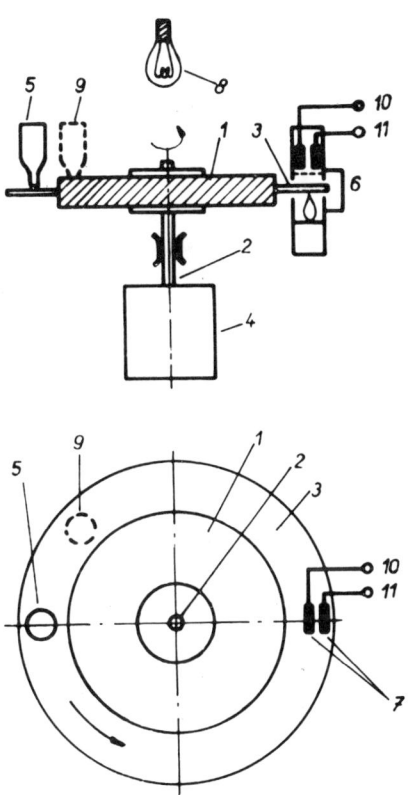

1, Rotating body; 2, shaft; 3, net; 4, motor; 5, column; 6, flame ionization detector; 7, electrodes; 8, infrared lamp; 9, solution for impregnation of the net; 10, 11, connectors to the amplifier

system needs to be simplified to render it more reliable and easier to operate.

The flame ionization detector responds to the mass entering it per unit time. It is a common misconception among chromatographers that, because the moving wire detector incorporates a flame ionization detector, the overall system is also a mass sensitive device. However, the mass entering the detector per unit time is related to the mass of solute brought into the system on the carrier. At a constant carrier speed, the film of column eluent on the carrier will be constant and, thus, the mass of solute entering the detector will be proportional to the concentration of solute in the eluent film coating the wire. The wire transport detector is thus a concentration sensitive device.

The Disc Detector

Dubský (10) developed a transport detector, employing a rotating gauze disc as the carrier. A diagram of his apparatus is shown in figure 7. It consists of a rotating disc, the perimeter of which is made of wire gauze. The column exit is situated just above the gauze, and the effluent flows through the gauze, the excess being collected below the gauze. A little ahead of the point of coating, in the direction of rotation of the disc, is situated an infrared lamp, which evaporates the solvent, leaving the solute coated on the gauze. Diametrically opposite to the point of coating is situated the flame ionization detector. The flame jet is situated beneath the gauze, such that the flame itself is in contact with the gauze. The electrodes of the flame ionization detector are placed above the gauze directly over the flame. The ions collected by the electrodes provide a current, which is fed to an amplifier and thence to a recorder. The system has the advantage of being simple, compared with the conventional wire detector, but although the authors claim a tenfold increase in sensitivity, this is difficult to confirm, as they do not specify sensitivity in terms that permit comparisons with other detectors. Szakasito and Robinson (11) claimed that the metal gauze disc carrier produced excessive noise and, in particular, "spikes", resulting from local concentrations of solute accumulating during evaporation at the intersections of the wire mesh. The wire disc was replaced by an alumina disc 4.5 in. in diameter and with the edge tapered to 0.25 mm thick. This thin edge was used for coating and detection. It was claimed that a significant reduction in noise was achieved, but the sensitivities obtained were not specified in terms that would allow comparison with other detectors. The life of the alumina disc seems questionable, as it would seem that, in continued use, the pores of the alumina would eventually become blocked by incompletely combusted solutes, or mobile phase components, such as buffers or inorganic substances, which would not burn. The disc does appear a more simple transport system than the wire or chain, but its reliability and sensitivity still remains to be established.

REFERENCES

1. A. T. James, J. R. Ravenhill and R.P.W. Scott, Chem. Ind., (1964) 746.
2. E.O.A. Haahti and T. Nikkari, Acta. Chem. Scand., 17 (1963) 2565.
3. A. Karmen, Anal. Chem., 38 (1966) 286.
4. R.P.W. Scott and J. F. Lawrence, J. Chromatogr. Sci., 8 (1970) 65.
5. L. M. Van Dijk, J. Chromatogr. Sci., 10 (1972) 31.
6. A. Stolyhwo, O. S. Privett and W. L. Erdahl, J. Chromatogr. Sci., 11 (1973) 263.
7. V. Pretorius and J. F. van Rensburg, J. Chromatogr. Sci., 11 (1973) 355.
8. K. Šlais and M. Krejčí, J. Chromatogr., 91 (1974) 181.
9. N. A. Dugger, U. S. Patent, 528,343.
10. H. Dubský, J. Chromatogr., 71 (1972) 395.
11. J. J. Szakasito and R. E. Robinson, Anal. Chem., 46 (1974) 1648.

PART 4

THE USE OF DETECTORS IN LIQUID CHROMATOGRAPHY

CHAPTER 1

The Selection of the Appropriate Detector

There are six commercially available detectors for general use from which the liquid chromatographer can choose for any particular application. They are namely, the refractive index detector, the conductivity detector, the UV detector, the fluorescence detector, the wire transport detector and the electron capture detector. The most commonly used detector is the UV detector followed in popularity by the refractive index detector. These two detectors are the "work horse" detectors of liquid chromatography, the others being chosen for very specific applications. No single detector can perform as a universal detector capable of handling all types of separation problems, but in the author's experience, the UV detector has the greatest versatility, which, combined with high sensitivity and wide linear dynamic range, should, therefore, be the liquid chromatographers first choice in the selection of his armory of detectors.

The UV Detector

The UV detector can only be employed to detect those substances that adsorb in the UV range of wavelengths but, as already discussed, the majority of substances do adsorb to some extent in this range. However, if the liquid chromatographer does have some specific application in mind, he should check that the substances of interest adsorb UV light at an appropriate wavelength. The UV detector has a fairly high sensitivity and, thus, can be used for the detection of trace components in a mixture and because it has a relatively wide linear dynamic range, it can be used for both qualitative and quantitative analysis. Its relatively high sensitivity in general makes it unsuitable for detection in preparative chromatography, but if a variable wavelength UV detector is available, a wavelength can often be chosen where the solutes exhibit relatively little adsorption, thus reducing its sensitivity and making it more appropriate for detection in preparative work. The UV detector, however, does not lend itself to solvent gradient elution as those solvents necessary for use in gradient elution adsorb to different extents in the UV and, thus, during gradient development, the base line is irregular or exhibits serious drift. Rabel (1) suggested a series of solvents for gradient development, that would be used with the UV detector that minimized the drift and base line irregularities but does not eliminate them. Although in general, the UV detector is at a disadvantage when used with gradient elution development, in ion exchange chromatography, however, changes in buffer concentration frequently do not affect the base signal of the UV detector anything like the extent that solvent changes do. Thus the detector can often be used effectively under gradient elution development in ion exchange chromatography. Modern UV detectors have small cell volumes and small diameter connecting tubes, and so they can be effectively used with modern high efficiency microparticulate columns without impairing the resolution obtained. UV detectors having high sensitivity , operating at one or perhaps two fixed wavelengths are available at prices ranging from $1500 to $2000. Unless the chromatographic analysis is to be restricted to very specific applications, the UV detector is recommended as the first choice for general chromatographic work.

The Refractive Index Detector

The refractive index detector, being a bulk property detector, has a significantly lower sensitivity than the UV detector but more nearly approaches the universal detector in the sense that it will detect any solute that has a refractive index different from that of the mobile phase. The presence of substances that do have the same refractive index as the mobile phase can easily be checked by changing the refractive index of the mobile phase without changing significantly the polarity characteristics of the mobile phase. Thus the sequence of elution of the solutes will remain the same. For example, if the mobile phase consists of a specific mixture of hexane and ethyl acetate, the hydrocarbon component can be replaced by heptane or octane, and the mixture rechromatographed. The solvent mixture will now have a different refractive index, and any solute that had the same refractive index as the hexane/ethyl acetate mixture will be detected without significantly changing the elution order of the solutes. The refractive index detector, as well as being less sensitive than the UV detector, has a smaller linear dynamic range, perhaps only two orders of magnitude and is thus less suitable for the quantitative analysis of a mixture having solutes present over a wide concentration range. Due to its relatively low sensitivity it is commonly used as the detector in preparative liquid chromatography and can be recommended for such use. It also finds a major application in the detection of solutes that do not have UV chromaphores. It is often used as the detector in gel permeation or, more correctly, exclusion chromatography, particularly in the separation of hydrocarbon polymers, carbohydrates and polypeptides, although polypeptides can be detected with the UV detector. The refractive index detector is even less suitable for gradient elution development than the UV detector and is not even suitable for use with gradient elution in ion exchange chromatography. The refractive index detector is a robust detector, simple to operate and is relatively inexpensive, costing in the neighborhood of $2,000. It is recommended as the second choice in liquid chromatography detectors.

The Fluorescence Detector

The remaining four commercially available liquid chromatographic detectors will now be discussed, but an order of preference will not be given. The choice of these detectors will depend solely on the nature of the mixture the chromatographer is required to separate. The fluoresence detector is very specific, detecting only those substances that fluoresce, but, as a result of its specificity, it can have an extremely high sensitivity. Although it has a very high sensitivity, it has a poor linear response and has a linear dynamic range which may be less than one order. For quantitative work, it is strongly recommended that the detector is calibrated against known standards over the complete concentration range that it is to be used. Its major area of application, besides the detection of those substances that naturally fluoresce, is for the detection of primary amines, in particular, amino acids and peptides employing an appropriate fluorescing reagent such as Fluoropa or Fluorescamine. This necessitates the use of a post column reactor which must be carefully designed to minimize band dispersion. Fluorescence detectors are available at a range of prices, but an effective and reliable detector can be obtained at a cost between $1800 and $2500. If the fluorescence detector is purchased combined with a UV detector, the cost of the dual system can be considerably more. The fluorescence detector is recommended for use where solutes naturally exhibit fluorescence or for the detection of substances containing primary amine groups, such as peptides in conjunction with an appropriate fluorescing reagent.

The Wire Transport Detector

The wire transport detector, although a solute property detector, has a sensitivity more in keeping with a bulk property detector i.e. about 2×10^{-6} g/ml. It is probably the nearest approach to the universal detector, in that it detects all substances containing carbon. It has a linear dynamic range of over four orders, but this includes solute concentrations that are much in excess of those normally used in liquid chromatography. Thus, the linear dynamic range in practice for the wire transport detector, is only about three orders of magnitude. The great advantage of this detector is that it imposes no limits on the nature of the solvents used for the mobile phase, the only restraint being that the solvents are reasonably volatile. It is, therefore, very appropriate for use with gradient elution development. Employing the solvent series developed by Scott and Kucera (2) for gradient elution, a gradient program can be carried out scanning a solvent polarity range from heptane to water. As the series of twelve solvents range widely in refractive index and contain many solvents that adsorb strongly in the UV, the wire transport detector is the only commercially available detector that can be used. Thus, the detector can be employed in the separation of multicomponent mixtures of wide polarity range, such as reaction mixtures, cosmetic products, such as lipstick, creams, oils, lotions, etc., providing a chromatogram of all the components in a single gradient elution development procedure. The wire transport detector, however, has a number of disadvantages, other than its relatively low sensitivity. The detector is a complex, bulky instrument and costs about $6,000, which is expensive compared with other liquid chromatography detectors. Due to its complexity, it requires more knowledge, skill and experience to operate it successfully. In the commercially available detector, significant band dispersion occurs in the detector system. The dispersion, however, is sufficiently low enough to be tolerated even when using high efficiency column and occurs not in the mobile phase transport system where dispersion is almost negligible but in the subsequent gas lines from the oxidizing furnace to the flame ionization detector. Despite these drawbacks, however, the detector has to be recommended for use with gradient elution development and, particularly, in the separation of multicompound mixtures containing solutes having widely different polarities.

The Electron Capture Detector

The electron capture detector is a relatively recent development and can, on the one hand, exhibit an extremely high sensitivity but on the other is exceedingly specific in its response. As the detector has only been introduced recently, confirmation of its performance in analytical liquid chromatography laboratories is somewhat sparse. The detector has a restricted linear dynamic range of less than two orders of magnitude and should be carefully calibrated for accurate quantitative work. As a result of the vaporizing system, the detector produces some band dispersion, but this can usually be tolerated even when using high efficiency columns, in order to realize the detector's high specific sensitivity. The detector's specific response to electron capturing substances such as the halogenated compounds used as pesticides and fungicides makes it very useful for pollution monitoring and control and, in some cases, for the detection of carcinogenic compounds. In fact, at present, its major area of application is in the identification and detection of such components. The cost of this instrument is about $4500.

The Electrical Conductivity Detector

The electrical conductivity detector is another sensitive specific detector that has special areas of application. The detector can only be used to detect ionic substances and thus, the mobile phase must be an ionizing medium, and this usually takes the form of aqueous solutions sometimes containing a polar solvent, together with an appropriate buffer. The mobile phase itself can have a reasonably high conductivity, as the detector can function at reasonably high sensitivities even in the presence of a significant base signal but is unsuitable for gradient elution involving significant changes in buffer concentration. The detector can be made very small in volume and thus, band dispersion in the conductivity cell can be made minimal and not impair the performance of high efficiency microparticulate columns. This detector can be recommended for use in detecting ionic materials and can be purchased at a cost of $1500 - $2500.

In choosing an appropriate liquid chromatography detector, the UV detector and the refractive index detector can be recommended for general work, where a wide range of different solute mixtures are required to be separated and analyzed. The other detectors should only be chosen because their functional characteristics specifically suit the type of samples that are to be analyzed. A versatile liquid chromatography laboratory would employ both UV and refractive index detectors and have at least two other specific detectors available if required.

REFERENCES

1. F. M. Rabel, Amer. Lab., 6 (1974) 33.
2. R.P.W. Scott and P. Kucera, Anal. Chem., 45 (1973) 749.

Chapter 2

Quantitative and Qualitative Analysis

The technique of liquid chromatography has developed rapidly over the past decade, and because of the introduction of high sensitivity, linear detectors is now not only being used as a separation technique, but also for quantitative and qualitative analysis. The chromatographic data provided by the detector output gives the retention time or retention volume of a peak, by which the solute may be identified. From the relative peak height or peak areas, the proportion of the solute originally present in the mixture can also be determined. However, the precision of the chromatographic data obtained depends not only on the performance of the detector, but on the control of the chromatographic conditions and also the method of measuring the data. Scott and Reese (1) investigated the overall precision that could be obtained from liquid chromatographic data and determined the necessary control over the chromatographic variables to obtain retention data with a precision of 0.1%.

Solvent Composition

The composition of the solvent used as the mobile phase can have a profound effect on solute retention and is used as an operating variable to control the retention of the solutes in a given mixture. It follows that if retention times are required to have a precision of 0.1%, then the solvent composition must be maintained sufficiently constant to maintain the required precision. In figure 1, curves relating the corrected retention times of three solutes are plotted against solvent composition. Solvent composition is taken as the % w/v of the polar solvent in the dispersing solvent. In fact, it has been shown (2) that the reciprocal of the retention volume is linearly related to the solvent composition, but over a small solvent composition range, the relationship can be taken as linear. Thus the points in figure 1 are force-fitted to a linear function and the results are summarized in Table I. It is seen that, to achieve a precision

Figure 1

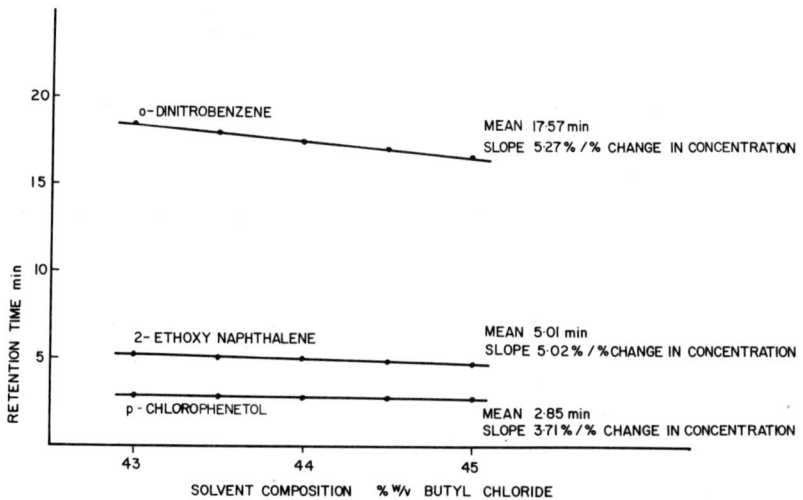

Table I

Solvent Concentration Tolerances for Retention Time Precision

Solvent	Retention Time at 44% of Butyl Chloride in Heptane, min	Tolerance for Retention Time Precision	
		Concentration Tolerance for 1% Precision, % v/v	Concentration Tolerance for 0.1% Precision, % v/v
p-Chlorophenatol	2.85	±0.14	±0.015
2-Methoxynaphthalene	5.01	±0.10	±0.010
o-Dinitrobenzene	17.27	±0.10	±0.010

of 0.1%, the solvent concentration must be maintained to within 0.01% w/v. This level of control of solvent composition is fairly easy to maintain providing a closed solvent system is employed, but it is extremely difficult, if not impossible, to make up a solution to this accuracy using volatile solvents. It is, therefore, recommended that large bulks of solvent are made up if precise results are required, and each new solvent checked by chromatographing a standard solute. Appropriate correction factors can then be calculated and employed where retention times are to be compared with previously used solvents.

The Solvent Pump

Scott and Reese employed the Walters 6000M pump for their work, which was found to give a flow of mobile phase over a period of about 12 hours, with a standard deviation of 0.07%. This performance was amazingly good but such precision can only be maintained if the pump is operated with the necessary precautions. The majority of chromatographers treat precision liquid chromatography pumps as just another piece of plumbing hardware, whereas they should be treated with the care and respect given to an analytical balance. It was found that, to maintain the precision, the following procedures needed to be taken. The pump should never be allowed to run dry, otherwise abrasion between piston and cylinder produces small leaks. Any mobile phase used should be filtered through a 0.2 micron millipore filter, particularly if the solvent had been dried over activated silica gel or alumina. The usual method of filtering by a filter paper was found to be inadequate. The filter contained in the pump should be regularly changed, and the inlet tubes to the pump should have as large a bore as possible (2 mm i.d.) to prevent the pump being starved of solvent. The pump should never be subjected to a back pressure greater than the rated maximum.

The mobile phase must be brought to a constant and fixed temperature, and thus constant density, prior to entering the column and, if the volume flow rate through the column is to be maintained constant, then the pump must deliver a constant mass flow rate to the column. As the pump is designed to provide a constant volume flow rate, then it must be supplied with solvent at a constant density and its displaced volume must also remain constant. If a precision of 0.10% is required, then the displaced volume of the pump and the solvent density must also be maintained constant to within this level of precision. It is almost impossible to thermostat the pump so the ambient temperature of the pump and solvent reservoir must be controlled, and this means that the temperature of the room in which the apparatus is situated must be controlled. Most pumps are made of stainless steel, which has a coefficient of cubical expansion of about $1.3 \times 10^{-5}/°C$ i.e., $0.0013\%/°C$ and thus, the effect of ambient temperature changes on the pump volume will be negligible. The cubical expansion of solvents, however, is much higher and for heptane is $1.25 \times 10^{-3}\%/°C$ i.e. $0.125\%/°C$. Thus, to maintain the solvent density of 0.1%, the ambient temperature must be maintained constant to $0.4°C$. This control of ambient temperature is not unreasonable in normal heat-controlled and air-conditioned laboratories, but has to be maintained if the required precision is to be achieved.

Column Temperature

It is well-known that the retention volume and retention time of a solute varies considerably with temperature, and in figure 2 the retention volume of the three solvents determined over a narrow temperature range is

Figure 2

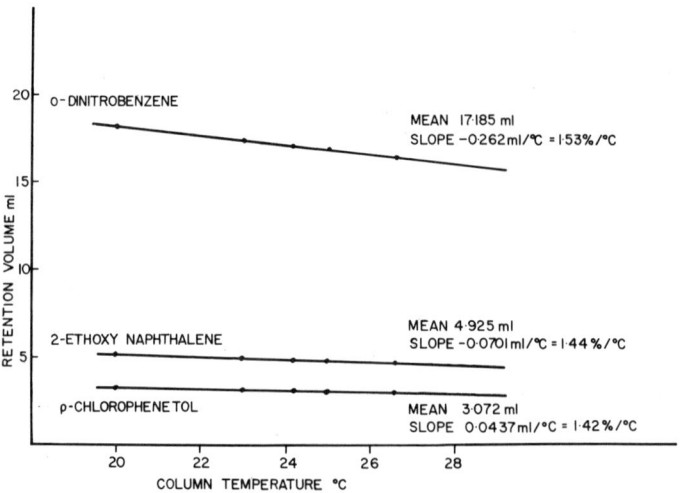

GRAPHS OF CORRECTED RETENTION VOLUME AGAINST COLUMN TEMPERATURE FOR THREE SOLUTES

shown plotted against temperature. The relationship between retention volume and temperature is, in fact, logarithmic, but over the small range of temperatures concerned, it is approximately linear, so a linear function is force-fitted to the results for the three solvents. A summary of the results obtained from the regression analysis of the data used in figure 2 is shown in Table II. It is seen that, to attain a precision of a 0.1%, the

Table II

Temperature Tolerances for Retention Time Precision

Solute	v' at 23.8°C	k'	Temperature Control for 1% Precision	Temperature Control for 0.1% Precision
P-Chlorophenatol	3.072 ml	0.945	±0.35°C	±0.04°C
2-Methylnapthalene	4.925	1.519	±0.35°C	±0.04°C
o-Dinitrobenzene	17.185	5.301	±0.33°C	±0.03°C

temperature of the solvent and column must be maintained to within 0.04°C. It is not difficult to maintain this level of temperature control on the thermostat bath, but it can be extremely difficult to return to a given temperature to within 0.04% (after prior change). It should also be pointed out that column temperature control to 0.04% would be extremely difficult, if not impossible, to obtain if an air bath was employed. Due to the relatively low thermal capacity and specific heat of air, local variations within an oven of 1°C can usually be found in gas chromatograph hot air ovens. Thus, for precise work, liquids are recommended as the thermostating medium for liquid chromatographic columns.

Sample Load

The mass of the sample injected onto a liquid chromatographic column can significantly affect both the solute retention time and column efficiency (2, 3). In figure 3, the retention times obtained from a column for the solutes o-dinitrobenzene and 2-ethoxy naphthalene are shown, plotted against mass of solute injected into the column. It is seen that, for precise comparative work, either the mass of sample injected must be kept constant, or the total mass of each solute maintained at a level below 0.1 µg. Thus, irrespective of the specifications of the detector and its

Figure 3

Graphs of Retention Time Against Sample Mass for Two Different Solutes

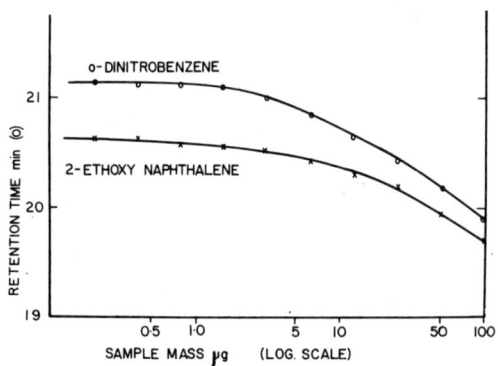

performance, for precise results, the operating conditions of the chromatograph as a whole must be carefully controlled. However, the method of measuring the data can also affect the precision of the measurement obtained. There are generally two methods of data measurement--manual and automatic data processing involving, at one extreme, a simple electronic integrator, and at the other, computer processing.

Manual Measurement of Chromatographic Data

Retention distances can be measured manually on the recorder chart with a good quality plastic rule. All distances should be estimated to the nearest 0.1 mm. The base line under each peak is constructed, using a sharply tipped pencil and the peak height taken as the distance between the constructed base line and the center of the recorder trace at the peak maximum. The peak width is taken as the distance between the inside edge of the recorder trace on the leading edge of the peak to the outside edge of the recorder trace on the trailing edge of the peak at 0.6065 of the peak height. The peak widths are best measured by means of a 3X comparitor and graticule which should be calibrated in units of 0.2 mm or less. Peak widths can then be estimated to the nearest 0.1 mm. The peak area can be taken as the product of the peak height and the peak width.

Computer Data Processing

There are two general methods of data acquisition by a computer; in the first, the output from the detector is sampled, a limited amount of data is temporarily stored; the data is processed, and the processed data in the form of retention times, efficiencies, peak areas, etc. is permanently stored; in the second method, the data is sampled, and each data point in the chromatogram is permanently stored, the chromatographic data required is obtained by subsequently processing the stored data. The former method is inflexible and, as the data is discarded, reprocessing by an alternative procedure is not possible, and a reconstructed chromatogram cannot be obtained. The second method is far more flexible, the data can be processed by any chosen method or by a number of different methods, if required, and as the raw chromatographic data is permanently available, a reconstructed chromatogram, in whole or in part, can be obtained on any chosen scale and presented on a cathode ray tube screen or plotter. The stored data can also be statistically tested, and any points representing outliers can be rejected.

There are two main factors in the acquisition of data that can affect the precision of the chromatographic results obtained; they are the data sampling rate and signal noise level from the detector. Minicomputers that are employed for chromatographic data acquisition have, in general, a maximum sampling rate of 240 samples/sec, which has to be shared between the number of stations involved in the time-sharing system, and there will be a maximum sampling rate for any one station of 60 samples/ sec. The limit of 240 samples/sec is imposed on the computer by the auto ranging amplifiers associated with the A/D convertor that is necessary to utilize the complete linear dynamic range of the detector. If a disc storage system is used in conjunction with the computer, the time of data transfer to the disc will also limit the sampling rate of the computer but not reduce it below 240 samples/sec. The 60 sample/sec limit for each station results from the fact that a sampling rate in excess of the 60 Hz frequency of the main electricity supply would result in unacceptable noise. Thus, if there are 10 stations from which data is to be acquired, then the maximum sampling rate for each station will be 24 data points/sec. The fact that a finite data acquisition rate exists causes a discrimination limit to be imposed on any chromatographic results obtained, unless the data is processed from the disc using special software. Consider a peak having a time width of 10.3 sec sampled at a rate of 5 samples/sec, then a value of either 10.2 or 10.4 will be taken by the computer as the peak width. Thus, assuming the peak is eluted at a retention time of 360 sec, and employing the standard equation for column efficiency:

$$N_1 = 4 \times (360/10.2)^2 = 4893$$

$$N_2 = 4 \times (360 \times 10.4)^2 = 4793$$

Thus, the sampling rate will permit a discrimination of 190 theoretical plates in 4983, equivalent to 3.0%, and this precision of measurement, due to the sample acquisition rate being 5 samples/sec, cannot be improved. It can be shown in a similar way that, irrespective of the control over chromatographic conditions, a precision of 1% in column efficiency cannot be realized unless the data acquisition rate is greater than 20 samples/ sec.

Figure 4

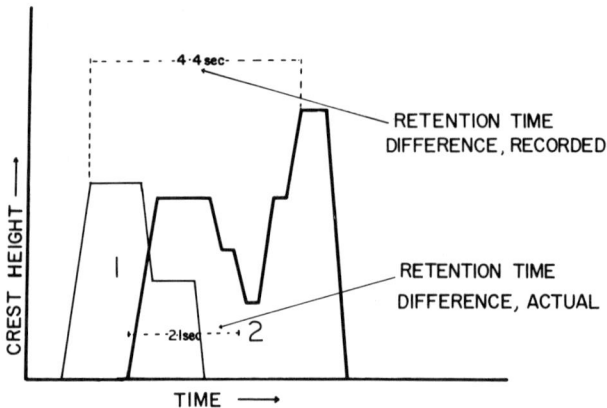

PEAK CRESTS RECONSTRUCTED BY THE COMPUTER
(99.9–100% PEAK HEIGHT)

The noise level of the signal that is digitized can also significantly affect the overall precision of the chromatographic data obtained. An example of this is shown by the crests of the two replicate peaks, shown in figure 4, taken from computer data and expanded on an oscilloscope screen. It is seen that, although the differences in retention time between the two peaks is only 2.1 sec, as a result of a noise spike on the front of the crest of the first peak and the noise spike on the back of the peak crest from the second peak, the measured difference in retention times is 4.4 sec. It follows that, to attain the highest precision, the noise has to be significantly reduced or, if possible, eliminated from the detector signal.

There are several ways of reducing noise. The first and obvious method is to average a number of points and take the average value for data acquisition. This procedure is frequently used and is sometimes called the "slice" method of data acquisition. This method has the advantage of reducing noise without distorting the peak, but at the same time may reduce the data acquisition rate which is also undesirable. Another method of reducing noise is to use an "on the fly", exponential smoothing procedure. This procedure is very effective, does not reduce the data acquisition rate, but tends to distort the peak. If all data points are stored on disc, very sophisticated smoothing procedures are possible, including the rejection of outliers; furthermore, this procedure in no way affects the rate of data acquisition. An alternative procedure is to interpose a filter circuit between the detector output and the A/D convertor. This procedure has already been discussed and, again, does not affect the rate of data acquisition but, if not designed correctly, can produce peak dispersion and asymmetry. Further an active filter would be more effective than a passive filter, as the active filter would have a much sharper frequency cutoff and, thus, provide more efficient noise rejection.

Qualitative Analysis

The basic chromatographic measurements for qualitative analysis are retention times (T_R) and retention volumes (V_R). This requires the accurate location in time of the peak maximum which makes small demands on the detector. The response of the detector need not be linear, although as previously discussed, the signal should be as free from noise as possible. Although T_R and V_R are the basic chromatographic measurement, they are not normally employed for solute identification purposes.

From the theory of chromatography, the retention volume V_R for a liquid solid system is given by:

$$V_{R(A)} = V_o + K_A A_S$$

when V_o is the void or dead volume of the column
K_A is the distribution coefficient of the solute A
and A_S is the surface area of the stationary phase.

It should be noted that K_A is the characteristic that will permit the identification of the solute. Now both V_o and A_S will vary from column to column, depending on its packing density and, thus, the variability of V_o is eliminated by chromatographing a completely non-adsorbed solute N, which will have a retention volume equivalent to V_o. Thus, the corrected retention volume V'_R is given by

$$V'_{R(A)} = V_{R(A)} - V_o = V_{R(A)} - V_N = K_A A_S$$

The only variable left to eliminate is A_S, and this can be eliminated by chromatographing another standard solute S added to the original mixture containing Solute A,

then, $$V'_{R(S)} = V_{R(S)} - V_o = V_{R(S)} - V_N = K_S A_S.$$

thus, $$V'_{R(A)}/V'_{R(S)} = (V_{R(A)} - V_o)/(V_{R(S)} - V_o) = (V_{R(A)} - V_N)/(V_{R(S)} - V_N)$$

$$= K_A A_S / K_S A_S = K_A / K_S$$

The function $V'_{R(A)}/V'_{R(S)}$ is called the retention ratio of solute A to the standard solute S and is the value that permits the identification of solute A to be confirmed as its value depends only on K_A and K_S both of which are solely characteristic of the solute and not dependent on the packing characteristics of the column.

At a constant flow rate, all retention volume measurements can be replaced by retention times.

Thus, $$T'_{R(A)}/T'_{R(S)} = (T_{R(A)} - T_o)/(T_{R(S)} - T_o) = (T_{R(A)} - T_N)/(T_{R(S)} - T_N) =$$

$$K_A A_S / K_S A_S = K_A / K_S$$

Retention times can be measured extremely accurately under carefully

Table III

Precision of Retention Time and Retention Ratio Measurements

	1	2	3
k'	0.94	1.50	5.21
Mean Retention Time (min)	6.283	8.119	20.421
Standard Deviation (sec)	0.38	0.20	0.46
Standard Deviation (%)	0.10	0.04	0.04

	T_2/T_1	T_3/T_2
Retention Ratios Mean	1.2922	2.5153
Standard Deviation	0.00119	0.00111
Standard Deviation (%)	0.092	0.044

where T_1, T_2 and T_3 are the retention times of solutes 1, 2 and 3 respectively.

controlled chromatographic conditions. In Table III, the precision of retention time and retention ratio measurements determined by Scott and Reese are given, and it is seen that the standard deviations of retention times are about 0.1% or less. The precision of retention ratios are about the same, but the retention ratio data can be used for different columns having different packing characteristics provided the same stationary phase is used in conjunction with the same solvent system. It is not advisable to rely on retention ratios obtained from one phase system to unambiguously identify an unknown substance. The retention ratios of the known and unknown solute to a given standard should be compared on at least two phase systems. If the known and unknown substances give the same retention ratios to the standard respectively on each phase system, then more confidence can be placed in the identification of the unknown solute.

Quantitative Analysis

Quantitative analysis by liquid chromatography, as opposed to qualitative analysis imposes stringent demands on the performance of the detector. The detector must have a linear response and must be operated within its linear dynamic range; further, the base line noise must be minimal if peak area measurements are to be employed. The basic measurements employed for quantitative analysis are peak heights or peak areas. In general, analysis by peak heights give more precise results than peak areas. A simple form of the procedure is as follows. A known weight (W_A) of the solute to be determined (A) is chromatographed with a known weight (W_S) of standard S. Let the peaks for solute A and standard S have peak heights of h_A and h_S respectively.

Then,
$$\frac{W_A}{W_S} = \alpha \frac{h_A}{h_S} \quad \text{or} \quad \alpha = \frac{W_A h_S}{W_S h_A} \quad (1)$$

where α is known as the calibration constant.

A known weight W'_S of the standard is now added to the mixture containing an unknown weight W'_A of solute A and a sample of the mixture chromatographed. If in the resulting chromatogram, the solute peak and the peak for the standard have peak heights of h'_A and h'_S respectively, then:

$$\frac{W'_A}{W'_S} = \alpha \frac{h'_A}{h'_S} \quad (2)$$

After substituting α from equation 1, the unknown weight of solute A in the mixture is given by:

$$W'_A = \frac{W_A h_S}{W_S h_A} \times \frac{W'_S h'_A}{h'_S} \quad (3)$$

This procedure can be used employing peak areas instead of peak heights by substituting values for peak areas for peak heights in equation 3.

Equation 3 is quite general and can be used for any number of different substances in the mixture (providing they are adequately resolved) using only one standard, and the calibration factor is determined for each different solute relative to the standard.

If the detector response is the same for all solutes which is sometimes the case, particularly when separating polymers using a refractive index detector, then a normalization procedure can be used. Under these circumstances, if the peak heights are h_1, h_2 h_3....h_n for solutes 1, 2, 3....n, then the percentage of any solute P is given

$$\text{by} \quad \% P = \frac{h_p}{h_1 + h_2 + h_3...h_n} \times 100 \quad (4)$$

Again, peak areas can be substituted for peak heights in equation (4) if desired. The precision of quantitative analytical data employing peak

Table IV

Precision for the Analyses of the Mixture of Normalization by Peak Heights and Peak Areas Measured by Computer with a 0.5 sec Input-Time Constant

		Peak 1	Peak 2	Peak 3
	k'	0.94	1.50	5.21
Analysis by Peak Heights				
Mean		1.937	16.491	81.574
Standard Deviation		0.0465	0.121	0.148
Standard Deviation (% of the mean)		2.46	0.736	0.18
Analysis by Peak Area				
Mean		0.633	7.486	91.884
Standard Deviation		0.032	0.072	0.0823
Standard Deviation (% of the mean)		5.071	0.97	0.09

heights and peak areas as determined by Scott and Reese and obtained under carefully controlled chromatographic conditions is given in Table IV. It is seen, except for peaks eluted, late in the chromatogram peak height analysis gives more precise results than peak area analysis. The standard deviations for components determined at levels of 2.0%, 10% and 82% are 0.05%, 0.12%, and 0.15% respectively, demonstrating that precise results can be obtained from liquid chromatographic analysis providing sensitive, linear and stable detectors are employed. It should also be emphasized that the precision given above is not solely dependent on the characteristics of the detector but are equally dependent on the careful control of the chromatographic conditions.

REFERENCES

1. R.P.W. Scott and C. Reese, J. Chromatog., 138 (1977) 283.
2. R. E. Majors, Anal. Chem., 44 (1972) 1722.
3. R.P.W. Scott and P. Kucera, J. Chromatogr. Sci., 12 (1974) 473.

CHAPTER 3

Practical Hints on Detector Operation

The basic functions and operational procedures for using liquid chromatographic detectors are normally given in the manual supplied with the equipment. However, in the operation of the apparatus for practical liquid chromatographic analyses, the instructions given by the manufacturers are sometimes inadequate. The following points may be of value to the practicing chromatographer to help him realize the full potential of his equipment.

The Recorder

Before commencing work, the recorder should be zeroed preferably by adjustment of the offset control to 10% of full scale deflection. Adjusting the zero at 10% F.S.D. permits some negative drift of the base line without the trace going off scale. Recorder sensitivity should then be adjusted and there may be on a recorder a single damping control, a single sensitivity control, or both. If only single controls are available, these should be increased until a transient signal on the recorder will produce an immediate deflection without "overshoot" or balancing delay. If two controls are available on the recorder, adjust the damping control to about 20% of its full value and then increase the sensitivity control until a transient deflection again comes rapidly to the balancing point without "overshoot". If this cannot be achieved without the pen deflecting in excess of the balance point, then the damping control should be increased until the "overshoot" is eliminated. The correct adjustment of the sensitivity and damping controls are essential otherwise at high detector sensitivities there will be significant short term noise on the base line or alternatively the response of the detector will be slow and sluggish resulting in stepped peaks.

The Amplifier

Most liquid chromatography detectors have a meter displaying the amplifier base signal and the base signal should be adjusted within the correct operating range given in the instrument manual. With the base signal within this range the amplifier zero control should be adjusted to give a zero deflection on the recorder at 10% F.S.D. During operation if the amplifier zero drifts then it should be brought back to position by the amplifier zero control and not by the recorder offset zero control. If a zero position cannot be obtained by adjustment of the amplifier, then the cells are out of balance and should be brought into balance by one of the methods described below. It is common among chromatographers to operate their detectors at as high a sensitivity as possible. In practice it is better to operate the detector at the minimum sensitivity that is practical in order to provide the maximum stability of the detector system. The sensitivity used will be governed by the maximum charge that can be placed on the column without impairing its efficiency. If trace analysis is required, it may be necessary to operate the detector at its maximum sensitivity but initially the detector sensitivity should be set at mid-range and then adjusted to meet the requirements of the specific analysis concerned.

Detector Cells

The UV detector and the refractive index detector have sample cells and reference cells. To obtain a balanced output, the two cells must provide similar signals to the photocells and amplifier, therefore the reference cell must be filled with the mobile phase being used with the column. If the column eluent is changed, then the reference cell must be filled with the new mobile phase. Some workers when using nonadsorbing mobile phases such as heptane do not fill the reference cell but obtain a balance when it contains air only. Although a balance can often be obtained in this way, this is an undesirable procedure as the offset signal due to the heptane in the sample cell, which is balanced off by the zero adjustment, encroaches on the linear range of the amplifier and may reduce it significantly.

If the two cells when filled with the same mobile phase still do not produce a balance, then it is likely that the cell windows are contaminated and require to be cleaned. In reassembling the cell after cleaning, care should be taken not to tighten the cell locking screws too tightly otherwise the cell windows or lenses will be damaged. However, the cells must be tightened sufficiently so as not to permit leaks.

Bubbles -- Their Removal and Prevention

Bubbles can occur in the cell producing a violent off scale deflection that can be either negative or positive. If the problem is in the reference cell, the deflection is usually negative and conversely in the sample cell the deflection is positive. Bubbles can form by air dissolved in the mobile phase particularly when aqueous mixtures such as methanol and water are used. It is a good practice to degas all solvents prior to use by shaking in a flask under vacuum before filling the solvent reservoir. To aid in the elimination of bubbles should they form, the column should be connected to the lower connecting tube entering the detector and exit from the upper connecting tube via another tube bent downwards in the form of a U. This permits a slight negative pressure to be applied to the cell aiding in the removal of bubbles and as any bubble entering the bottom of the cell will rise, it will pass readily through the cell and exit tube. Connecting the column to the upper connecting tube of the detecting cell has the opposite effect and makes bubbles extremely difficult to remove.

If a bubble remains in the cell, then the mobile phase flow rate should be increased and a slight back pressure applied to the exit of the cell by restricting the flow. This reduces the size of the bubble and as a result of the increased pressure, often permits its free exit. If the bubble is particularly recalcitrant, then the union between the column and the detector should be loosened and a flow of polar solvent such as acetone forced back through the cell causing the bubble to be removed via the loose union. One note of warning should be added. When applying back pressure to the cell to remove bubbles, the pressure should not be allowed to become so great as to cause leaks between the cell window and the gaskets.

Spurious Peaks

During the development of a chromatogram, unexpected and foreign peaks are sometimes observed. There can be a number of causes for these peaks.

1. Elution of Sample Solvent

In some cases it is necessary to dissolve a sample prior to injection in a solvent that does not have the same composition as the mobile phase. Under such circumstances the solvent employed in the sample may be eluted as a peak sometimes in the same chromatogram or sometimes as a broad peak in a subsequent chromatogram. The only solution to this problem is to use the mobile phase as a solvent for the sample.

2. Contaminated Syringe

If the syringe has not been cleaned adequately from the previous sample, then small peaks from the original sample will result, interfering with the chromatogram of the fresh sample. Considerable care, therefore, should be taken to clean syringes and contamination is particularly likely when using 1 μl syringes.

3. Contaminated Septum

    After continual use, the septum in an injection system can absorb many of the solutes previously used for analysis and in time these will bleed from the septum particularly when an injection is made. The solution to this problem is to replace the septum.

4. Air Dissolved in the Sample

    Air dissolved in the sample will usually be eluted close to the dead volume and will change the refractive index of the mobile phase. This change in refractive index will produce a peak on both the refractive index detector and the UV detector, particularly when using high detector sensitivities. To eliminate this effect, the sample should also be degassed.

5. Detector Displacement Effects

    When using polar solvent in the mobile phase, the quartz windows of some detector cells adsorb a thin layer of the polar solvent on the surface. If a more polar solute is eluted from column, in some cases it will displace the polar solvent and this will result in a change of light absorption and produce a spurious peak at the beginning of the eluted peak. Subsequent to the solute passing from the cell, the polar solute again replaces the absorbed solute producing a positive or negative tail to the peak. This effect can be reduced or eliminated by choosing a more polar solvent for the mobile phase but at the same time reducing its concentration so the elution sequence of the solutes remain sensibly constant.

Base Line Instability

Detector noise in its various forms has already been discussed but in practice measures can often be taken to reduce or eliminate them.

Short Term Noise

Short term noise appears as "grass" on the base line and this type of noise can normally be completely eliminated. If the short term noise persists at low detector sensitivities, the source is usually in the recorder and can be eliminated by either slightly reducing sensitivity or increasing the degree of damping. If the short term noise only occurs at high sensitivity, then this is arising from the detector amplifier and can be eliminated by interposing an active or passive filter between the amplifier output and recorder. The design of the filter circuit is discussed in Part I, Chapter 5, page 45. Another source of short term noise arises from the pulsation from a reciprocating pump. This can be usually identified by the fact that the frequency of the noise matches the frequency of the pump stroke. Pump noise can be reduced by incorporating a pulse damping device between the pump and the column and also by thermostating the detector cells. The effect of thermostating detector cells has been discussed in Part 3, Chapter 2, page 110.

Long Term Noise

Long term noise as already defined has a frequency of the same order as the eluted peaks and thus is far more difficult to reduce than short term noise or drift. In fact, it is the long term noise that ultimately limits the sensitivity of the detector. Long term noise can increase progressively during the lifetime of a chromatographic column as it becomes contaminated by trace materials from the samples analyzed which accumulate on the stationary phase and eventually elute irregularly. A silica column can often be cleaned by the following method: six column dead volumes of heptane, dichloromethane, ethyl acetate, acetone, ethanol and water are passed sequentially through the column. This procedure completely deactivates the column and in the process may wash out the majority of the contaminating impurities. The column then has to be reactivated and this can be achieved using the same solvents and the same volumes of each solvent and passing them through the column in the reverse order. This procedure should significantly reduce any short term noise that has developed and if it does not then the only alternative is to replace the column with a new one.

Drift

There are two main sources of drift both due to non-equilibrium conditions in the column and detector. If the detector column and mobile phase is not in thermal equilibrium, then serious drift can result. This can be eliminated by thermostating the detector column and mobile phase supply. Another source of drift results from incomplete mobile phase equilibrium with the stationary phase. Such drift always occurs on changing the composition of the mobile phase and to eliminate this drift, mobile phase should be pumped through the column detector system until a stable base line is obtained.

The Moving Wire Detector

The moving wire detector requires more skill and experience for satisfactory operation than any other commercially available detector. Noise can arise in many parts of the equipment but one of the major sources comes from impurities in the gases employed. For this reason the gases should be carefully cleaned using adsorption tubes that remove both traces of carbon dioxide and hydrocarbon gases. Absorption tubes packed with a suitable alkali and Linde Molecular Sieves are appropriate and one of each type of absorption tube should be inserted in the nitrogen, hydrogen and air lines respectively. A satisfactory base current from the ionization detector should be less than 10^{-9} amps. The instrument noise should first be checked with the flame ionization detector extinguished and the wire stationary. Under these circumstances only amplifier noise will be apparent and this can be eliminated by the methods previously suggested. The flame ionization detector should then be ignited, the ovens and furnaces turned on and left for three hours to come into equilibrium. The standing current should then be checked to ensure that the gases are not contaminated. The older version of instrument contained a Linde Molecular Sieve absorption container fitted in the back of the detector. This initially cleaned the gases but rapidly became saturated and provided a reservoir of contamination that would continually bleed hydrocarbon gases into the gas stream. This absorption device should either be regularly renewed or if the gases are cleaned prior to entering the detector it can be completely removed and bypassed. Having tested the amplifier and FID system, the wire can now be turned on and any increase in noise noted. In general, the wire should only double the noise level observed on the recorder. If noise exceeds this amount, then the alignment of the wire through the various apertures should be carefully checked and adjusted. After running instrument for some time, spikes sometimes appear on the recorder trace between and during the elution of a peak. The spikes are usually due to contamination of the coating block or the coating block pulley. These spikes can be removed by washing the coating block and pulley with copious amounts of solvent. Under some circumstances spikes will only appear on peaks or the peaks will have a jagged appearance. This effect is usually due to poor wire coating and is often due to the nature of the wire. If the used wire spool is examined and the surface is mottled showing patches of brown or black together with gray or silver, than the wire spool should be changed for a new one.

The Conductivity Detector

The conductivity detector measures the conductivity of the mobile phase by determining the impedance between two electrodes situated appropriately in the column eluent. As discussed in Part 2, Chapter 4, this is achieved by making the conductivity cell one arm of a Wheatstone Bridge across which voltage is applied. It follows that only one, or in some cases neither, of the electrodes can be at earth potential. As the mobile phase is conducting there will be electrical continuity from the electrodes via the mobile phase to the column. There will also be electrical continuity between the column (usually made of stainless steel) by way of the connecting tubes to the pump which will be earthed. As a result of this indirect earth continuity to the electrodes an "earth loop" is formed which can cause serious detector instability. This instability can take the form of violent electronic noise at one extreme, and at the other, very sluggish movement of the recorder pen. This problem can be very difficult to deal with. The "earth loop" can interfere with either the operation of the recorder or the detector electronics. On some recorders there is a removable earth connection to the recorder imput. Removing this connection and "floating the recorder" may help to reduce this problem. In a similar manner some amplifiers can be floated with respect to earth. This should also be carried out wherever possible. The situation is exacerbated when using mobile phases of high conductivity. Mobile phases of high conductivity result from using buffers of high ionic strength. The earth loop can be reduced by employing significantly lower buffer concentrations and achieve approximately the same retention characteristics of the mobile phase by also adjusting the pH. In general changing the pH of the buffer has relatively little effect on the conductivity of the mobile phase whereas changes in buffer concentration can have a very significant effect on its conductivity. Some manufacturers attempt to reduce the "earth loop" condition by interposing a length of insulating tubing such as Teflon between the column and the detector cell. This in effect increases the impedance between the column and detector as it increases the path length of the mobile phase and helps reduce "earth loop" effects. This arrangement is satisfactory providing the insulated connecting tube has dimensions such that any band dispersion resulting from it is kept to a satisfactory minimum.

The Refractive Index Detector

The refractive index detector is very sensitive to fluctuations either in the pressure or the temperature of the mobile phase. If the pump employed is reciprocating in action, this often produces short term noise on the detector output. If the ambient temperature changes during a chromatogram than significant drift may also be experienced. The effect of pump pulses can be reduced by employing an effective pulse dampner. Thermostating the detector cell will eliminate thermal drift and also help reduce the detector susceptibility to changes in pressure. When using a refractive index detector it is a good practice to thermostat the cell and to use a single stroke piston type pump.

In some commercial refractive index detectors the optical system is designed to cover a specific range of refractive indices. Manufacturers of such instruments usually provide a number of cell or optical systems that are interchangeable and cover the practical range of refractive indices normally met in liquid chromatography. Where such instruments are used the refractive index of the mobile phase being employed should be checked and the appropriate optical or cell system used. If the incorrect optical system is used by mistake, it will usually be found that a balance point cannot be obtained with mobile phase in both cells or, if balance is obtained, the sensitivity of the system is extremely low.

Most commercial refractometers have a very restricted linear dynamic range, sometimes less than two orders of magnitude. When using the refractive index detector for quantitative analysis it is advisable to check the linear range of the instrument.

CHAPTER 4

Special Detector Techniques

Commercial detectors are designed to give a linear output that is directly related to the concentration of solute passing through them. The curve described by the recorder is thus a true representation of the Gaussian profile of the eluted peak. In the vast majority of liquid chromatography separations this type of output is required but there are a limited number of applications where an alternative form is desirable. All detectors that employ reference cells for comparison purposes can be connected to the column system in such a way as to provide alternative functional outputs. By appropriate but simple modification, detectors can be made to provide differential and integral outputs or display only changes in sample composition relative to that of a standard. In this chapter three examples of special detector techniques will be given but it should be emphasized that these technqiues will only be useful in certain limited areas of application and will be resorted to when the standard liquid chromatographic techniques are inadequate or unsuccessful.

The Differential Detector

The shape of the differential form of the Gaussian function has been discussed in Part 1, Chapter 2, page 6; it is sigmoid in shape with a positive maxima at the first point of inflexion of the Gaussian curve and a minimum at the second point of inflexion. If the peaks are completely resolved on the chromatogram then they can be clearly and easily identified in their differential form. If the peaks are not completely resolved, however, then the differential curve of the unresolved peaks are confused and extremely difficult to intepret. For this reason the differential form of the Gaussian function is very rarely used. If the elution profile of the solutes from the column are not Gaussian in form, however, then the differential form can be extremely valuable.

The concentration of a solute X_v eluted after v volumes of mobile phase have passed through the column will be some function of v, $(f(v))$.

Thus $X_v = f(v)$

and $\dfrac{dX_v}{dv} = \dfrac{d(f(v))}{dv} = f'(v)$

For finite values of dv, IE, Δv where Δv is small

then $\dfrac{\Delta X_v}{\Delta v} = f'(v)$

and $\dfrac{X_{v+\Delta v} - X_v}{\Delta v} = f'(v)$

or $X_{v-\Delta v} - X_v = \Delta v f'(v)$ \hfill (1)

Further if Δv is constant then $X_{v-\Delta v} - X_v$ will be directly proportional to $f'(v)$ the differential function of the eluted peak.

Figure 1

Detector Cells Connected to a Provide Differential Output

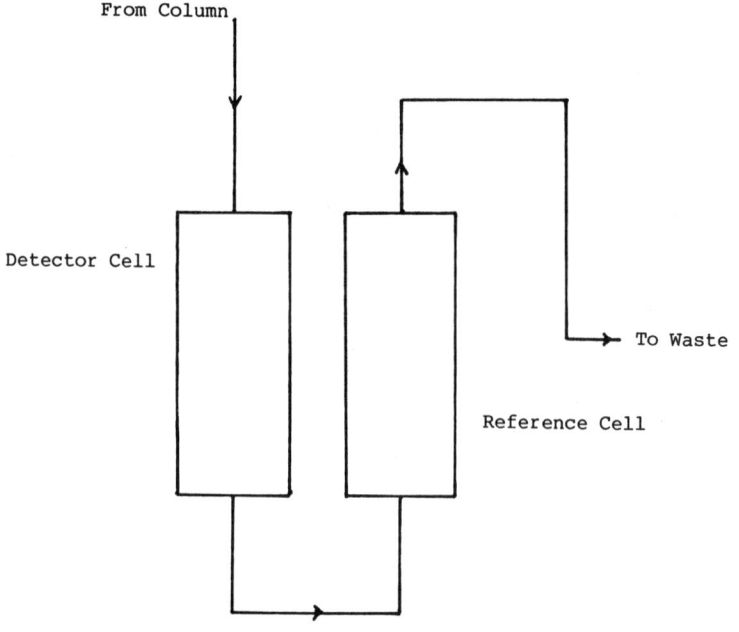

The value of $X_{v-\Delta v} - X_v$ can easily be measured in practice when using the UV detector or refractive index detector providing the detectors have reference cells. The column is connected to the detector in the manner shown in figure 1. This procedure was first employed in gas chromatography by Boeke (1) and was introduced to liquid chromatography by Essigmann and Catsimpoolas (2). The column eluent passes directly to the sample cell and thence to the reference cell by way of a short length of tubing, the volume

Figure 2

The Separation of a Mixture of Tetracyclines

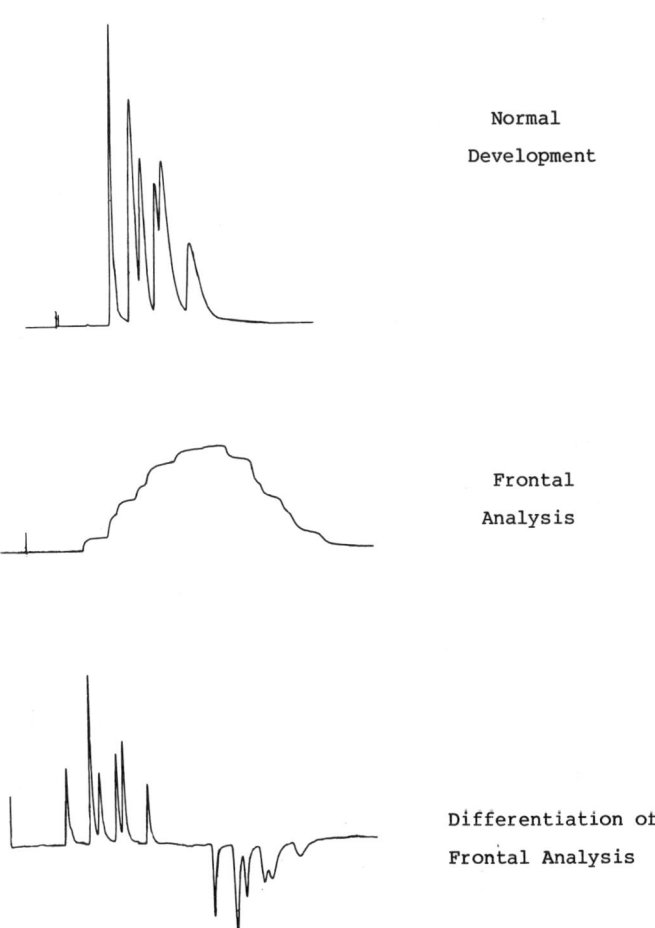

Normal Development

Frontal Analysis

Differentiation of Frontal Analysis

of which is the volume Δv. As the signal from the reference cell acts in opposition to the signal from the sample cell the output from the detector is $X_{v-\Delta v} - X_v$ as given in equation 1 and will always provide the differential of the function that describes the concentration of solute in the mobile phase.

An application of the differential mode of operating the UV detector is shown in the chromatogram at the top of figure 2. The chromatogram shows the separation of a mixture of different tetracyclines and it is seen that there is very serious peak tailing. This tailing results from the poor desorption characteristics of the solute from the stationary phase, silica gel. The quantitative results from such a chromatogram would give an extremely poor precision particularly if peak areas are employed. Seeking an alternative phase system that would affect a satisfactory separation could be a very time consuming operation. However, it should be noted that the front of each peak is relatively sharp and if the tails of the peaks could be made equally sharp then the separation could produce satisfactory analytical results. In the center chromatogram shown in figure 2 the mixture is separated by frontal analysis (3). This is achieved by employing a sample valve with a large volume loop (16 ml). The use of large volume sample loops for preparative chromatography and frontal analysis has been discussed by Scott and Kucera (4) and the resulting chromatogram shows firstly a series of rising steps as each solute breaks through the column and finally a series of descending steps as each solute is eluted from the column. It should be noted that the rising steps have sharp fronts like the peaks in the upper chromatogram whereas the descending steps are broad and diffuse similar to the tails of the peaks eluted normally. In the lower chromatogram the effect of differentiating the ascending steps of the chromatogram obtained by frontal analysis is shown. The same column and phase system was employed as that used to obtain the upper chromatogram but the sample cell and reference cell was connected by a piece of stainless steel tubing 13 cm long 0.03 in. I.D. having a volume of 19 µl (Δv). The total sample volume employed was 16 ml. It is seen from the lower chromatogram a greatly improved separation is obtained that is quite satisfactory for quantitative analysis. It should be pointed out, however, that only the differential curves from the ascending steps of the frontal analysis curves should be used. The chromatogram obtained from differentiating the elution steps will be even worse than those from the original elution chromatogram. The peak heights of the differential

Figure 3

Calibration Curves Relating Peak Height to Mass of Sample from
the Differentiation of Frontal Analysis Curves

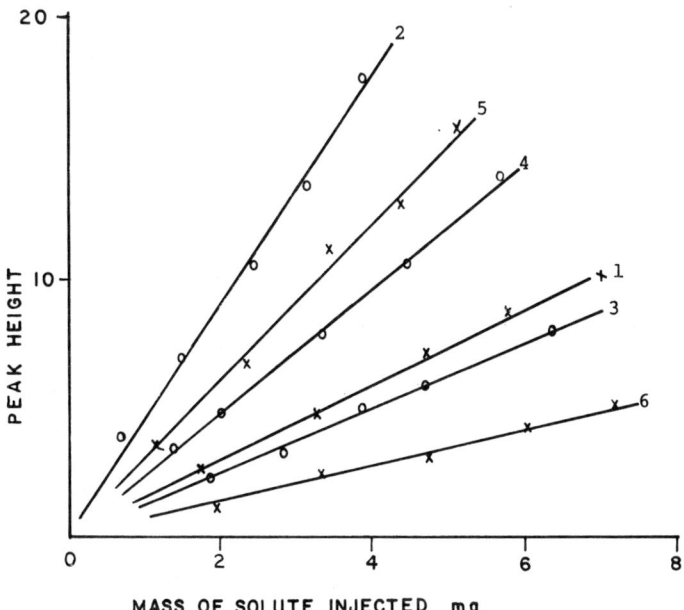

chromatogram are linearly related to sample mass as shown by the calibration curves given in figure 3 and thus can be used for quantitative analysis.

In figure 3 the peaks labeled one to six respectively refer to those substances shown separated in figure 2. Another example of the advantages of the differential mode of detector operation is shown in figure 4. In the upper chromatogram of figure 4 is shown the separation of an extract from a fermentation broth. It is seen that a very poor separation indeed, is obtained. In the lower chromatogram of figure 4 is shown the result of differentiating a frontal analysis of the same mixture. A good separation is obtained and again the chromatogram is quite suitable for quantitative analysis.

Figure 4

Chromatograms Demonstrating the Use of the Differential Display
in the Separation of a Fermentation Extract

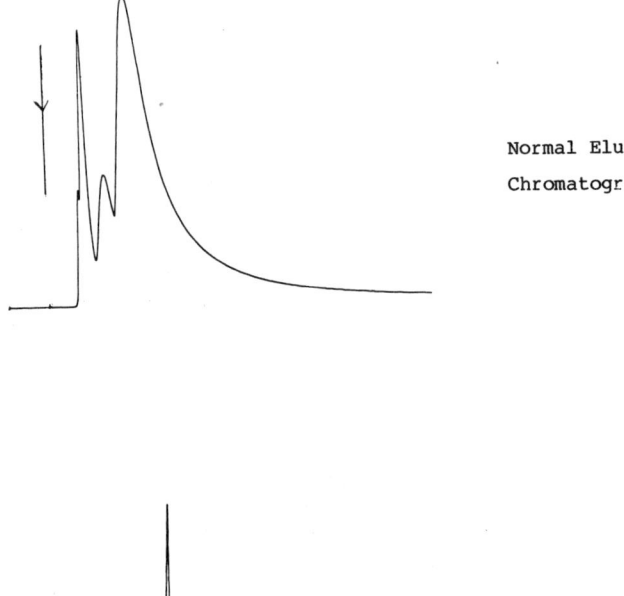

Normal Elution
Chromatogram

Differential Chromatogram
from Frontal Analysis

The use of the differential mode of detector operation can be extremely useful in cases where the normal chromatographic development gives very poor separation resulting from peak tailing. However, the technique does require significantly more sample for frontal analysis than with normal elution development so that sufficient sample must be available.

Integral Mode

Furthermore the response of the detector operating in its differential mode is two orders of magnitude less sensitive than when operated normally and thus high detector sensitivities have to be employed.

The Integral Detector

Any detector can be made to give an integral response if desired and for this purpose no special use of a reference cell is necessary. In practice the need for a detector having an integral response is very rare as electronic or digital integrators can carry out the same function more efficiently and with greater precision. However, for those who, for some reason or another, cannot utilize the more conventional methods of integration the alternative use of the detector as an integrator will be described.

If n solutes are completely eluted from the column in volume V

then $\quad \int X_1 dv \;+\; \int X_2 dv \;+\; \ldots \int X_n dv \;=\; M_1 + M_2 \ldots M_n$

where $X_1, X_2 \ldots X_n$ are the concentrations of the solutes $1, 2 \ldots n$ after a passage of mobile phase of volume v and $M_1, M_2 \ldots M_n$ are the masses of individual solutes in the mixture. Thus on the elution of solutes $1, 2 \ldots n$ the resulting integral curve will be in the form of a series of steps, the height of the

Figure 5

Experimental System to Provide Chromatographic Integration

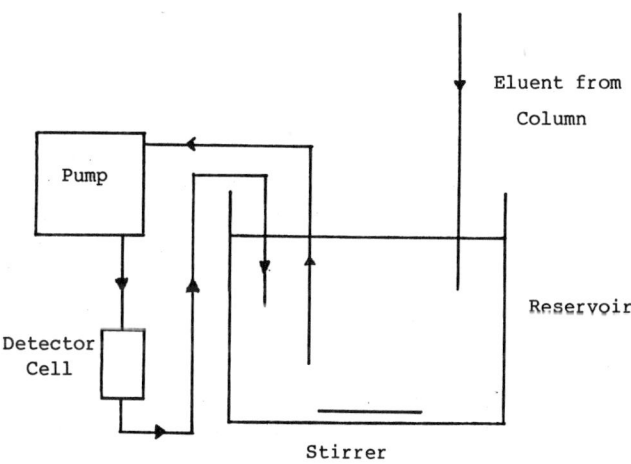

step for each solute being proportional to the mass present. It should be pointed out that analysis by normalization of each solute step can only be carried out if the detector response for each solute is the same. Alternatively the product of the step height and the response factor for each solute can be normalized to provide the required quantitative analysis.

Figure 6

Integral Curve Obtained by Chromatographic Detection

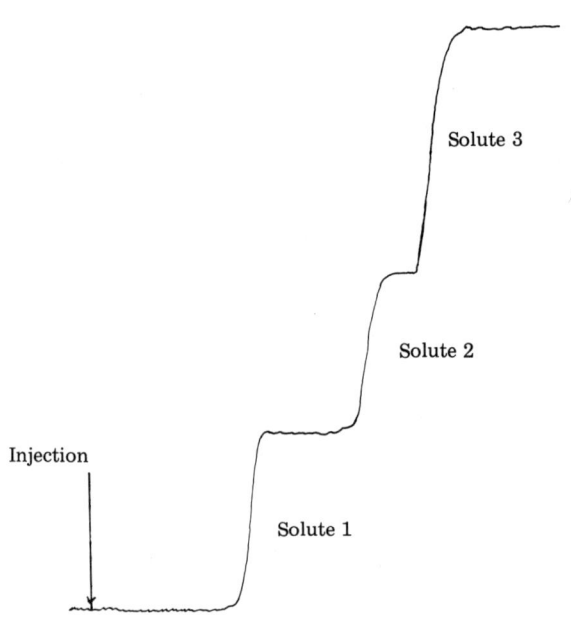

The system used to provide an integral of output from a detector is shown in figure 5. Mobile phase from a reservoir is pumped continuously through the sample cell of the detector and back again to the reservoir. The column eluent is allowed to flow directly into the reservoir and ideally the volume of mobile phase in the reservoir should be maintained constant. The volume of mobile phase in the reservoir should therefore be at least fifty times the volume of mobile phase through the column to completely elute all the solutes. If all the solutes are eluted in 10 ml of mobile phase then the volume of mobile phase in the reservoir should be at least 500 ml. The reservoir should be continuously stirred by suitable magnetic stirrer and the volume of mobile phase circulated through the detector and back to the reservoir should be maintained at about 1% of the total volume of mobile phase/min. In the above example a flow rate of 5 ml/min would be appropriate. An example of an integral chromatogram obtained by the above procedure is shown in figure 6. It is seen that a typical integral chromatogram is obtained that is quite suitable for quantitative analysis. As the total mass of each solute eluted from the column is in effect diluted by the 500 ml of mobile phase in the reservoir, high detector sensitivities need to be employed. Furthermore, it is advisable to use the maximum charge that can be placed on the column that will not impair the separation or the column performance. The reservoir can also act as a post column reactor. The appropriate reagent dissolved in the mobile phase in the reservoir will react immediately with any solutes eluted from the column to provide a derivative that can be detected by the particular detector being employed. Using the integral method of detection in this way can only be successful if the derivative of the solute is not labile. Many fluorescing reagents used in the detection of amino acids and peptides provide labile fluorescent products and so have to be used with some caution. In some instances the reaction conditions can be adjusted to give derivatives having reasonably permanent fluorescence.

Vacancy Chromatography

The term "vacancy chromatography" was introduced by Zhukhovitski and Turkel'taub (5) and is used to describe the chromatographic technique in which the mobile phase consists of a solvent containing a mixture of components which are maintained at constant concentration. Upon injection of a "sample", the resulting chromatogram may show negative and/or positive peaks depending on whether the components detected in the sample are present in lesser of greater concentration than they are in the mobile phase.

If mobile phase, carrying a constant concentration of solute X_o is fed continuously onto a chromatographic column and equilibrium is allowed to become established, the eluent from the column will also contain the solute at a concentration X_o. If a sample of the same solute dissolved in the mobile phase at a concentration X_1 is now injected onto the column where $X_1 \lessgtr X_0$ then this will produce a perturbation on the concentration X_o and from the plate theory, the equation for this perturbation when sensed by the detector at the end of the column will be

$$X_N = (X_1 - X_o) \frac{e^{-W^2/2N}}{\sqrt{2\pi N}} \qquad (2)$$

where X_N is the concentration of solute in the mobile phase in the Nth plate (i.e. that concentration sensed by the detector), v is the volume flow of mobile phase measured in plate volumes, N is the number of theoretical plates and W=v-N. It is obvious from equation 2 that if $X_1 > X_o$, a positive peak will be produced and if $X_1 < X_o$, a negative peak will be produced.

In the vast majority of chromatographic separations the sample is injected onto the column containing mobile phase only and thus the eluted peak is always positive representing an increase in solute concentrations in the detector. However the chromatographic process will operate in exactly the same way if the mobile phase contains a given concentration of the solute that is contained by the sample but the concentration of solute in the sample is less than that in the mobile phase. In this case on injection, a negative perturbation of solute in the mobile phase will be eluted through the column resulting in a negative peak being produced by the detector at the same retention time as a positive peak for the same solute would have been eluted. Such a system can be extremely useful for monitoring the composition of complex mixtures in process control. The apparatus used is shown in figure 7. A sample of the mixture of substances is chosen that has the composition that is required to be maintained in the process and dissolved in the mobile phase to a known concentration that is appropriate for the detector employed. The mobile phase is then pumped continuously through the sample valve column, detector and back to the reservoir. To monitor the process streams a sample is taken, diluted in mobile phase to the same concentration as the standard mixture in the mobile phase and injected onto the column. If the sample has the correct composition no peaks will be recorded by the detector. However, any substance present in excess of that in the standard mixture will show as a positive peak and any

Figure 7

Diagram of Apparatus for Vacancy Chromatography

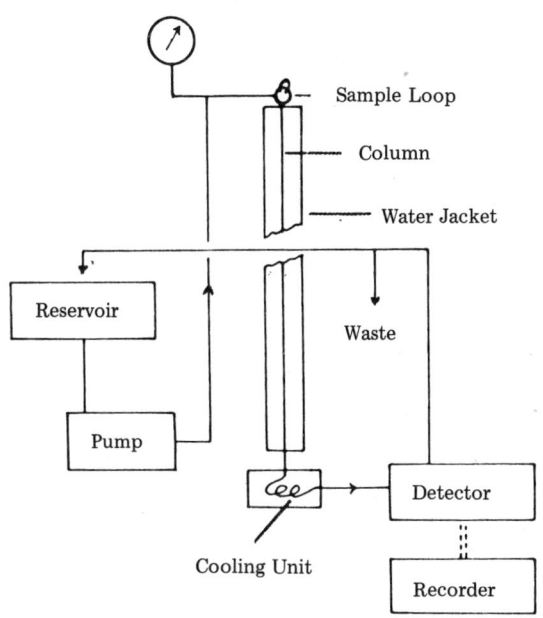

substance present below that in the standard mixture will be shown as a negative peak. The area of the positive or negative peak will be proportional to the excess or deficiency of the respective substance relative to the standard. Scott, Scott and Kucera (6) examined the technique of vacancy chromatography in the separation of nucleic acid bases and example of the use of the technique is shown in figure 8.

Figure 8a shows the positive peak obtained for a 600 µl injection of sample containing concentrations of uracil and hypoxanthine at the same levels as in the mobile phase but with the concentration of cytosine increased from 4 µg/ml to 4.4 µg/ml. There is no perturbation of the base line for the first three compounds because the mobile phase/stationary phase equilibrium is not disturbed by the injection. There is, however, a positive peak corresponding to the difference in concentration of 0.4 µg/ml

Figure 8

Chromatograms Obtained by Vacancy Development

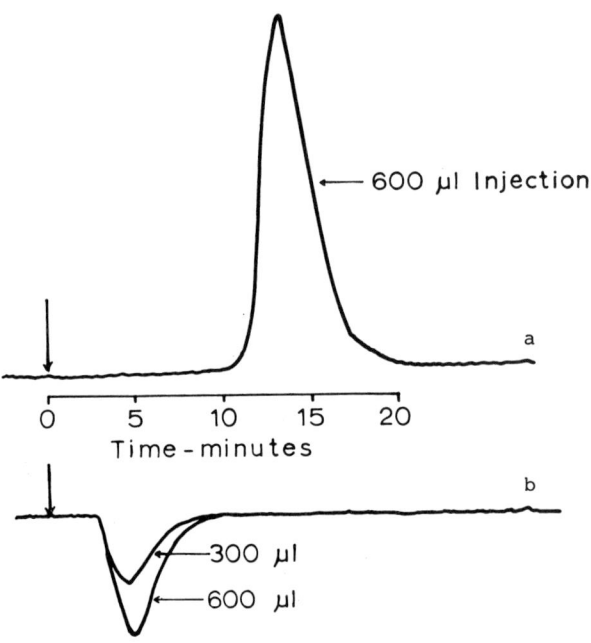

of cytosine. Figure 8b shows negative peaks obtained for 300- and 600-µl injections of sample equivalent to the mobile phase except for the omission of hypoxanthine. In this case a negative peak corresponding to the difference in concentration of hypoxanthine (0.5 µg/ml) is obtained. Under these conditions, differences in concentration of hypoxanthine can be determined without any interference from uracil from which, under normal elution conditions, it is not completely resolved.

Vacancy chromatography can be an extremely effective technique for identifying and determining changes in the composition of a complex mixture including the appearance of unknown impurities. This form of chromatography has, to date, not been exploited to any significant extent but it remains a very useful procedure for product or process control and is likely to become more popular as liquid chromatography increases its field of application.

REFERENCES

1. J. Boeke, Gas Chromatography 1960, (Ed. R.P.W. Scott), Butterworth, London, p. 88.
2. J. M. Essigmann and N. Catsimpoolas, J. Chromatogr., 103 (1975) 7.
3. R.P.W. Scott, Contemporary Liquid Chromatography, John Wiley and Sons, New York 1976, p. 11.
4. R.P.W. Scott and P. Kucera, J. Chromatogr., 119 (1976) 467.
5. A. A. Zhukhovitski and N. M. Turkel'taub, Dokl. Akad. Nauk. USSR, 143 (1961) 646.
6. R.P.W. Scott, C. G. Scott and P. Kucera, Anal. Chem., 44, (1972) 100.

CHAPTER 5

Spectroscopic Detectors

Modern liquid chromatography can separate very complex mixtures, but having achieved the required separation, there remains the problem of elucidating the structure of the eluted components. In the past the solutes eluted from the column were collected as fractions, concentrated and then examined by suitable spectroscopic methods. An alternative procedure is to pass the column eluent directly into a spectrometer and obtain the required spectroscopic data concurrent with the separation process. There are two ways to do this and the method used depends on the speed at which the spectroscopic data can be acquired and secondly on the type of spectroscopic data that is to be obtained. If the spectroscopic data can be obtained rapidly as in the case of the rapid scanning mass spectrometer the column eluent can be monitored continuously by passing the eluted solute via an appropriate interface device directly into the mass spectrometer and continuously monitor the column eluent during development. If the scanning rate of the spectrometer is relatively slow as in the case of the UV spectrometer then the interrupted elution procedure can be employed. The flow of mobile phase is stopped when the peak maximum has reached the absorption cell of the spectrometer, the scan initiated and when the spectrum has been obtained the flow of mobile phase is started again. Little or no loss in resolution is obtained by this procedure, as the only band dispersion that occurs mostly results from longitudinal diffusion, which in liquids is extremely small.

The five most common spectroscopic techniques for the elucidation of molecular structure are UV Spectroscopy, IR Spectroscopy, Raman Spectroscopy, Mass Spectometry and Nuclear Magnetic Resonance Spectrometry (NMR), the latter probably being the most informative spectroscopic technique. The NMR spectrometer, however, does not lend itself readily to on line monitoring of LC eluents. The technique is relatively insensitive requiring about 40 µg of sample to provide adequate spectra even in its most sensitive mode using Fourier Transform processing. Furthermore the volume of the sample cell is relatively large even when using microcells and would thus seriously impair the column performance. Coupling these disadvantages with the fact that the column eluent has to be passed through a spinning sample cell and proton free solvent would have to be employed, makes the NMR Spectrometer an unlikely instrument to couple with a liquid chromatograph and provide really useful spectroscopic data. The IR spectrometer suffers from similar disadvantages to that of the NMR spectrometer when considered on line with the liquid chromatograph. The IR spectrometer is relatively insensitive and even when used in the Fourier Transform mode of operation requires samples of the order of 10 µg to provide effective spectra. The cell volume is also relatively large and would seriously impair column performance. Further, most solvents normally used in liquid chromatographic development would interfere with the spectra obtained unless very restricted ranges of wavelength were scanned. A pulsed laser Raman system with Fourier Transform processing might be more feasible but such systems have yet to be developed. The UV spectrometer and mass spectrometer, however, do have adequate sensitivity, sufficiently small detector volume and when used with a suitable interface can be employed directly on line with the liquid chromatograph. Both systems are commercially available and have proved effective aids in the identification of solutes eluted from a liquid chromatograph.

The LC/UV Combination

The production of UV spectra from liquid chromatography eluents is usually carried out employing an interrupted elution procedure. The spectrometer itself is used to monitor the eluent at a fixed wavelength and at the peak maximum the mobile phase supply is stopped and the UV and/or visible spectrum scanned over an appropriate range of wavelengths. On the completion of the spectrum the mobile phase supply is again started and the development continued until the next peak is eluted. In this way a spectrum is obtained for each peak, or parts of a peak if resolution is incomplete, and if required the range of wavelength scanned can be chosen specifically for each respective peak. A continuously scanning UV spectrometer system for use with a liquid chromatographic column has been described by Denton et al. (1). These workers used an oscillating mirror rapid scanning spectrometer as the LC detector and demonstrated the advantages of recording the complete spectrum of each compound. However the flow cell was rather large (87 µl) which made it unsuitable for use with very high efficiency columns and the noise level was rather high, particularly below 250 nm.

Rapid scanning spectrometers with integrating array detectors such as Vidicons or solid state diode arrays are an alternative approach to LC/UV detectors and have been discussed by Talmi (2,3) and developed by Yates and Kuwana (4) and Milano, Lam and Grushka (5). A diagram of a diode array detector is shown in figure 1. The light source is focussed onto a slit S and onto a concave mirror M_1. Light then passes to an oscillating

Figure 1

The Diode Array Detector

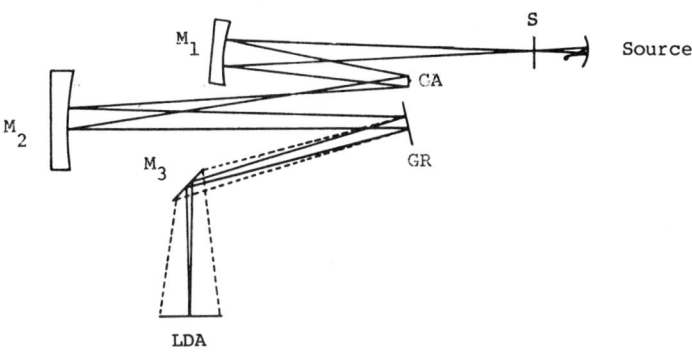

galvanometer GA where it is reflected onto another concave mirror M_2 and thence to grating GR. Light from the grating then passes to a third mirror M_3 where the slit is focussed on the diode array (LDA). In the system developed by Grushka et al. a deuterium lamp source was employed and the output from the diodes was interfaced with 16K core memory computer. Data was stored continuously during the development of the chromatogram and processed on completion. The total spectrum could be scanned over a period of 2.6 msec every 14 msec and 70 spectra averaged every second, but significantly lower scan speeds and scan frequencies were used in practice. The data stored during the development of a chromatogram can be subjected to smoothing procedures to improve the signal to noise ratio and thus increase sensitivity. As the data is permanently stored a range of smoothing methods can be tried and the best method that provides optimum smoothing chosen. A smoothed chromatogram obtained from the separation of a mixture of benzene, benzyl chloride and anisole and monitored at 254 nm is shown in figure 2. The presentation is taken from a cathode ray tube (CRT) screen and the indi-

Figure 2

Chromatogram Obtained from Diode Array Detector

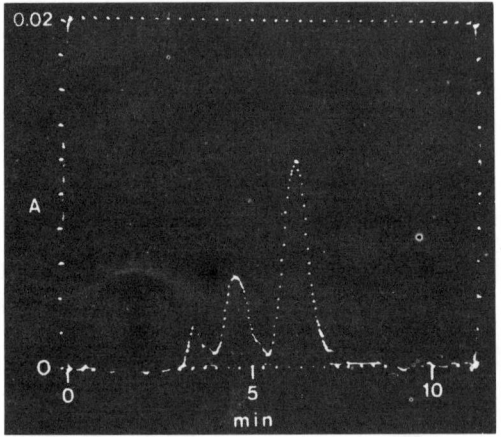

vidual smoothed points is clearly seen. In figure 3 is shown the spectra of each solute scanned at the peak maximum over the range 220 to 328 nm. The characteristics of the individual spectrum for each solute is clearly demonstrated. It should be pointed out that the spectra presented on the CRT gives a relatively poor impression of the quality of the spectra compared to presentation on a chart recorder. If the spectra shown in figure 3 were produced on a plotter then they would be seen to be completely satisfactory for structure elucidation. The diode array detector is not at present commercially available but the conventional UV scanning spectrometer modified for use with liquid chromatographs is manufactured by a number of instrument companies.

Figure 3

Spectra Obtained from Diode Array Detectors

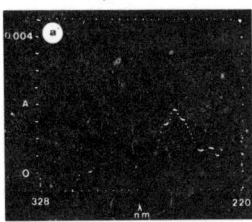 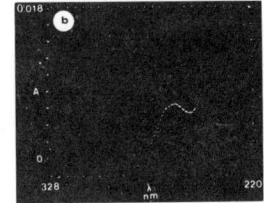 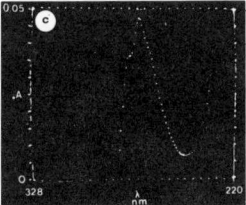

The Variscan LC/UV Spectrometer System

The Variscan is manufactured by Varian Associates. The output can be in the form of linear adsorbance units or percent transmission and tungsten-halogen or deuterium lamps are available as light sources to provide a wavelength range of 210 to 780 nm. Automatic wavelength scanning is available at scanning ratios of 10, 25, 50 and 100 per minute. The wavelength employed when using the instrument as a simple detector can be manually chosen and controls are fitted with coarse and fine adjustment. A diagram of the optical system is shown in figure 4. Light from the appropriate lamp passes through a Czerny-Turner monochromator fitted with selectable slits to one half of a dual chopper and thence to the sample compartment. After passing through the sample compartment the light passes through the second half of the dual chopper and is then sensed by a simple photo multiplier tube. The optical system is designed such that the optical paths through the sample and reference cells are symmetrical and contain an equal number of transmitting and reflecting elements. The two halves of the chopper are synchronized so that 100% of the light passes alternately through sample and reference cells. The cells are made of stainless steel, have a path length of 1 cm and a total volume (excluding connecting tubes) of 8 μl. The windows of the cells are made of quartz and the whole cell assembly can be thermostated.

An example of the use of the instrument to monitor the separation of a mixture of plant flavanones is shown in figure 5. The chromatogram on the

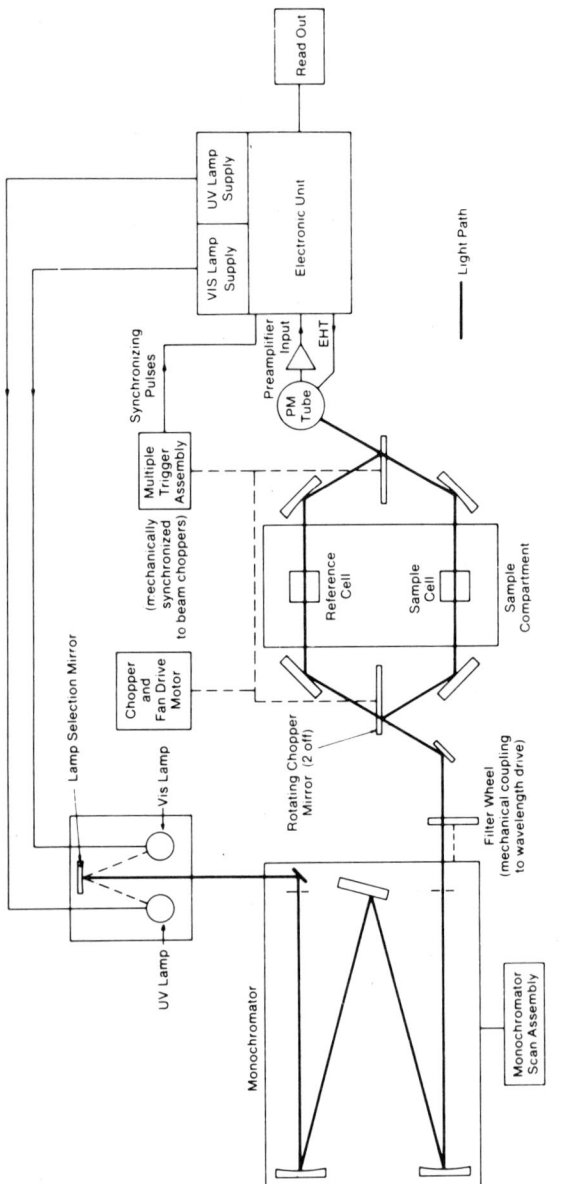

Figure 4

The Optical System of the Variscan LC/UV Spectrometer System

left depicting the separation was monitored at 310 nm which reduced solvent effects that arise from gradient elution development. The gradient used was from hexane to a 90/10% mixture of dichloromethane and isopropanol respec-

Figure 5

Chromatogram and Spectra of a Mixture of Plant Flavanones Obtained from the Variscan LC/UV Spectrometer System

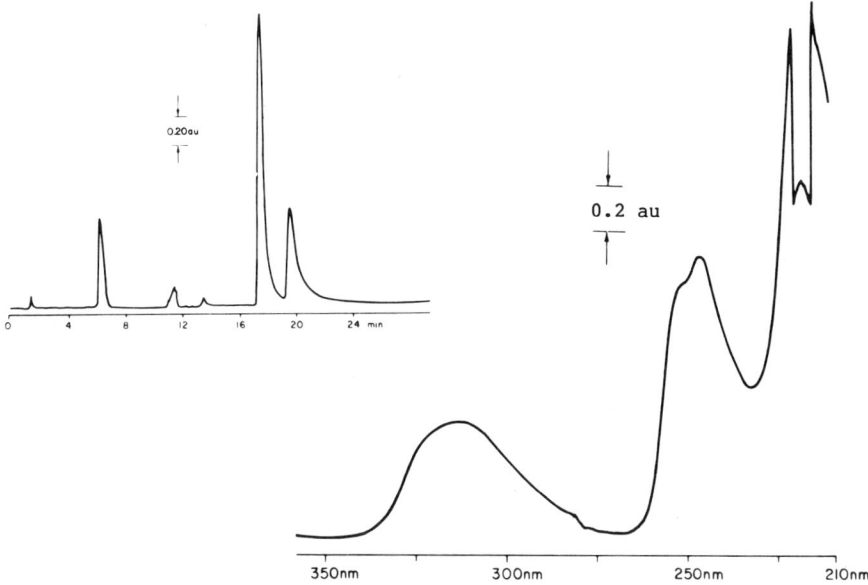

tively. The column was 50 cm long 2.4 mm I.D. packed with MicroPak silica gel. The spectrum shown on the right of figure 5 is for the first peak and it is seen that a very useful absorption spectra is obtained. The use of LC/UV systems for monitoring liquid chromatographic separations is simple and effective, unfortunately however, the value of the UV spectrum for structure elucidation is very limited.

The LC/MS Combination

Mass Spectroscopy is probably the ideal technique to combine with liquid chromatography as an aid to the structure elucidation of eluted components. Mass spectra can be obtained rapidly, requires sub-microgram amounts of material to provide satisfactory spectra and the data produced is highly informative with respect to molecular structure. There are two well established methods that can be used to interface a liquid chromatograph with a mass spectrometor. The direct inlet system developed by McLafferty et al. (6,7,8) and the wire transport system developed by Scott et al. (9,10). The former takes a proportion of the column eluent and passes it directly into a conventional mass spectrometer volatilizing both solvent and solute into the ion source. The latter employs the wire transport system in normal way; after passage through the column eluent stream, the solvent is evaporated from the wire and the solute coated on the wire passed through a suitable interface directly into the ion source of a quadruple mass spectrometer, where it is volatized into the electron beam. Both methods have been established as viable methods for LC/MS operation and the system involving the wire transport system is now commercially available. A third method reported by Horning et al. (11) is also a possible system for sample introduction from a liquid chromatograph into a mass spectrometer but has not yet been developed to its fullest capability. The method developed by Horning et al. involves the vaporization of a portion of the column eluent and both the solvent and solute vapor passes directly into a chamber containing a radioactive source. Positive ions are produced by a complex series of ion molecular reaction which then pass through a micropore aperture directly into the ion source of a quadrapole mass spectrometer. The authors claim a sensitivity of 5-10 ng per spectrum but it would appear that the solute must be reasonably volatile for the system to function. Furthermore, as the solvent is vaporized with the solute and both enter the mass spectrometer, the effect of gradient elution development (where the character and composition of the solvent is continuously changing) on the resulting spectra is uncertain.

LC/MS by Direct Sampling of Column Eluent

The direct inlet system for introducing a portion of the column eluent directly into a mass spectrometer was devised by McLafferty et al. (6,7,8) and a diagram of their interface is shown in figure 6. The Hitachi RMH-2 mass spectrometer was employed, modified to provide chemical ionization spectra. A 2500 l/sec differential pump was added to the ion source region to provide adequate removal of the solvent vapor. The column eluent passes into the ion source via a glass capillary (0.076 mm I.D.) which is contained along the center of a Teflon rod. The Teflon rod is inserted through the vacuum lock provided for probe injections and maintains a vacuum tight seal. The required restriction at the tip of the glass tube, to control the quantity of sample entering the mass spectrometer, was achieved by drawing out the tip in a small flame. At a flow rate of 10 μl/min through the glass

Figure 6

The LC/MS Direct Inlet System

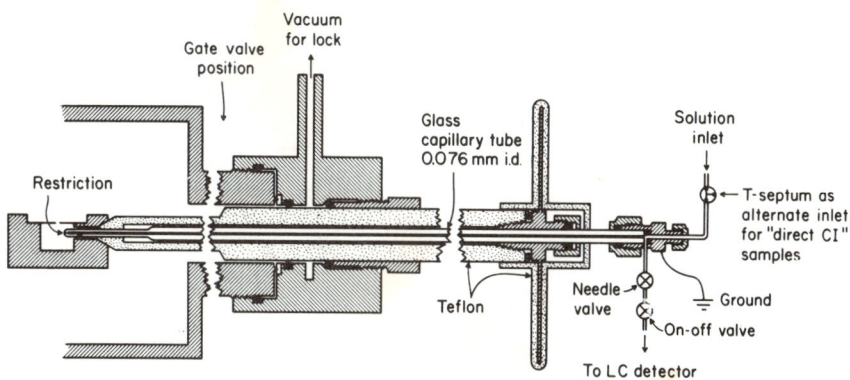

tube the delay time was about 6 sec. Flow rates in excess of 10 µl/min could be employed but the chance of electrical breakdown in the ion source was greatly increased. The authors state that if mobile phases having significant conductivity are employed electrical breakdown can be avoided by employing a quadrupole mass spectrometer as an alternative, where voltages at the ion source are closer to earth potential.

The solvent serves as the ionizing agent in the mass spectrometer and thus elution with mixed solvents or by gradient development can confuse the interpretation of the spectra produced. For example, 3-hexanone, benzene and methyl palmitate provide the protonated molecular ion MH^+ in methanol whereas the major peaks produced with pentane as the ionizing agent for 3-hexanone and androstanone result from hydride abstraction and are of the form $M-H^+$. The authors noted that compounds that have insufficient vapor

232   Liquid Chromatography Detectors

pressure for normal inlet procedures are not detected by this system, which unfortunately excludes a significant range of substances that are separated by liquid chromatography procedures. An example of chromatograms obtained by the McLafferty system is shown in figure 7. The liquid chromatograph was the Waters Associates ALC202 fitted with the M6000 pump for gradient

Figure 7

Chromatograms from the LC/MS Direct Inlet System

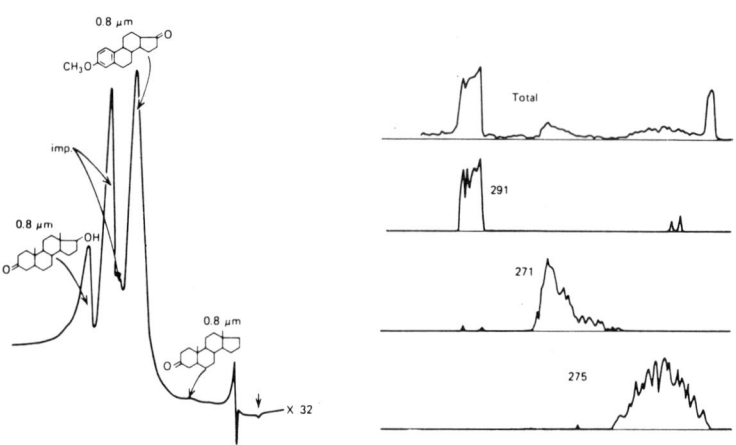

development. The chromatograms represent the separation of a mixture of S-α-3-androstanone, estrone methyl ether and androstanalone present at levels of approximately 200-250 μg. A linear gradient was employed over a period of 10 min, the initial solvent concentration being 40% acetonitrile and 60% water and the final solvent being 100% pure acetonitrile at a flow rate of 1 ml/min. The mass spectrometer was operated at a resolution of 1500, the source temperature was 200 °C, the ionizing electron energy 500 eV and the emission current 0.7 mA. The mass spectrometer scan speed was 10 sec/decade from mass 600 to mass 120 with a fly back time of 2 sec. The chromatogram on the left of figure 7 was obtained by a UV detector and the top chromatogram on the right obtained from the total ion current of the mass spectrometer. The three lower chromatograms on the right of figure 7 are reconstructed mass spectra using the ion masses of the respective quasi molecular ions of each component. It should be noted that the androstanone was hardly detected by the UV monitor but gave a significant peak on the chromatogram from the mass spectrometer. However the large peak shown by the UV monitor was not shown by the mass spectrometer. The authors claimed that 1 µg of cholesterol could be easily detected and 0.2 µg of tertiary butyl anthraquinone could be readily identified by monitoring solely on the peak mass of 265. The direct inlet system is obviously viable but has the disadvantage that it can only monitor substances that have significant volatility or vapor pressure at 200°C and furthermore the character of the spectrum produced will vary with the composition of the solvent employed.

The Wire Transport LC/MS System

This system, as already stated, utilizes the wire transport method of sampling the column eluent, by removing the solvent and passing the solute directly through the ion source of a quadrupole mass spectrometer where it is volatilized directly into the electron beam. The quadruple mass spectrometer is particularly applicable to this method of sample introduction, as the ion source is only a few volts above earth potential and the presence of an earthed wire passing through it does not interfere with the electron, or ion optics, of the electrode configuration. The success of the system hinges on the design of a suitable interface that can permit the passage of the wire from an environment at atmospheric pressure through the ion source and out again while maintaining a pressure of 10^{-6} mm of mercury in the ion source.

A diagram of the overall system is shown in figure 8. The wire employed was that supplied for the Wire Transport Detector, 0.005 in. O.D. and made of stainless steel. The wire from the drive system passes over an electrically insulated pulley, over a coating block (where the column eluent wets the wire), into the left hand interface of the mass spectrometer and thence through the ion source. It then exits through another identical interface, round another pulley and back to the drive system. In the lower portion of figure 8 is shown the location of the interface with respect to the ion source it is seen that the wire leaves the interface about 2 millimeters from the ion source and less than a centimeter from the electron beam. A potential is applied across the two pulleys causing a current of about 200 mA to pass through the wire. The heat generated by the current is rapidly conducted and convected from the wire in the air and thus the temperature of the wire only rises a few degrees above ambient temperature.

Figure 8

The Wire Transport LC/MS System

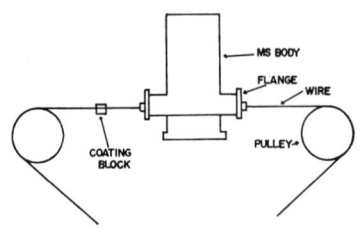

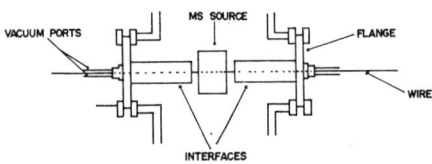

On entering the ion source, however, where the pressure is about 10^{-6} mm of mercury heat can only be lost by radiation and the wire heats up to a temperature of 200-300°C which volatilizes the adhering solute directly into the electron beam of the ion source. In this way substances having very low vapor pressure can be volatilized without decomposition and the system in effect acts as a continuous probe sampling device.

A diagram of the interface is shown in Fig. 9. The main body of the interface is constructed of stainless steel and is fitted to the side flanges of the Finnigan mass spectrometer, such that the interfaces are re-entrant to the ion source and terminate a few millimeters from the electron beam. The interface itself consists of two chambers separated and terminated by ruby jewels 1/10 in. in diameter and 0.018 in. thick. The jewels in the left-hand interface have central apertures 0.010 in. in diameter where the sample is introduced into the mass spectrometer. The jewels in the right-hand interface where the wire leaves the mass spectrometer to the winding spool have central apertures 0.007 in. in diameter. The larger

Figure 9

The LC/MS Wire Transport Interface

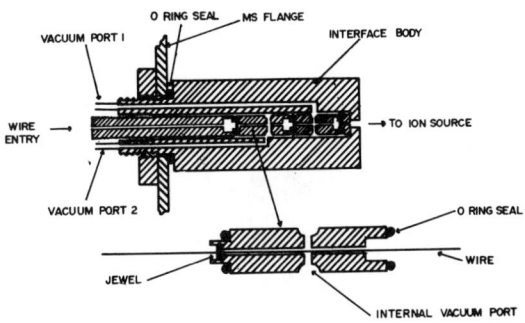

diameter apertures on the feed side of the spectrometer are employed to reduce scuffing of the wire and possibly loss of solute. The first chamber of each interface is connected directly to a 150 l/min rotary pump, which reduces the pressure in the first chamber to about 0.1 mm of mercury. The second chamber of each interface is connected to an oil diffusion pump backed by a 150 l/min rotary pump. The pressure in the second chamber of each interface was reduced by this system to about 5 to 10 μmHg. The entrance and the exit apertures were fitted with a helium purge T junction. Helium passed through the T junction replaced the air entering the mass spectrometer through the interfaces. In this way background spectra from air contaminants were reduced. The T junction also afforded a method of introducing methane or other suitable gases if chemical ionization spectra were required. A photograph of the apparatus is shown in figure 10.

The mass spectrometer can be operated under normal conditions except that the filament current employed should be about 3 mA. The instrument is used with the standard data handling system supplied for the Finnigan mass spectrometer and a typical set of operating conditions is given in Table I.

236   Liquid Chromatography Detectors

The sensitivity of the instrument for the diazepam was 4×10^{-6} g/ml in the column eluent. The wire samples the eluent at 10 µl/min at the maximum wire speed and thus as the scan speed of the mass spectrometer was 1 scan/sec this sensitivity corresponded to ca 7×10^{-10} g of diazepam per spectrum. The pressure in the source could be maintained at 1×10^{-6} mm of mercury. The use of helium as a purge gas into the interfacer reduces the noise level of the mass spectrometer by a factor of three with a commen-

Figure 10

The Wire Transport LC/MS Instrument

surate increase in sensitivity.

In figure 11 the chromatogram obtained from a fermentation extract is shown and this sample contains solutes that cover a wide range of polarity. The mixture was first chromatographed using the incremental gradient elution technique (12). A charge of 2 mg was injected onto the column and

Table I

Mass Spectrometer Operating Conditions

Filament current	3 mA
Electron energy	60 V
Source pressure	3×10^{-6} mm Hg
Mass range	60-199; 200-399; 400-650
Integration time	3; 5; 8
Sample/AMU	1; 1; 1
Threshold	1
Attenuation	5
Mass range setting	high
Mass run time	400 min
Delay between scans	3 sec

the detector used was the wire transport detector. The chromatogram produced is shown at the top of figure 11. During the chromatographic development fractions were collected for each peak and numbered as shown in the diagram. These were concentrated in a current of nitrogen to about 0.1 ml and each fraction examined using the LC/MS system as a probe injection device. A drop of each sample was spotted onto the wire for a few seconds and the total ion-current traces of these probe samples are shown on the right-hand side of figure 11. All the fractions were placed on the moving wire over a period of 8½ min. Thus the spectrum for an individual probe sample was obtained in approximately 25 sec. The sample was then chromatographed on the LC/MS system using a 6-mg charge and the same operating conditions. The chromatogram obtained from the total ion-current monitor is shown as a hard copy on the left-hand side of figure 11. It is seen that due to the nature of the hard copy presentation of the chromatogram it is somewhat compressed in size relative to the original using the wire transport detector. However, it is seen that the same pattern of peaks is obtained although owing to the larger charge size employed, the resolution is not as good.

Figure 11

Chromatogram of a Fermentation Extract by Incremental Gradient Elution (IGE)

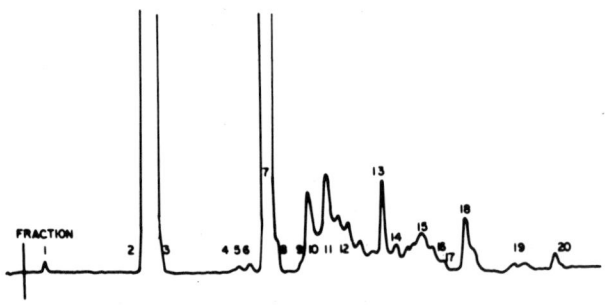

TOTAL ION CURRENT CHROMATOGRAM
BY LC/MS OF FERMENTATION EXTRACT

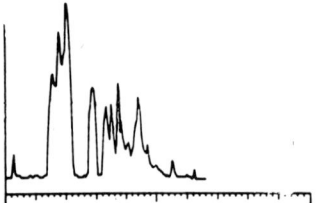

TOTAL ION CURRENT TRACE OF FRACTIONS
SAMPLED DIRECTLY ONTO WIRE

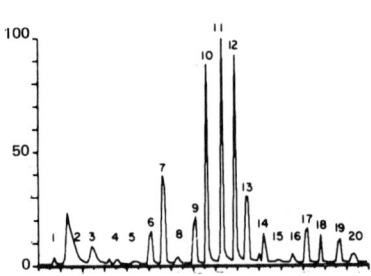

ION CURRENT TRACE FOR MASS 327 FROM
FRACTIONS SAMPLED DIRECTLY ONTO WIRE

ION CURRENT TRACE FOR MASS 327
FROM LC/MS CHROMATOGRAM

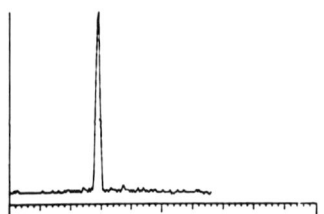

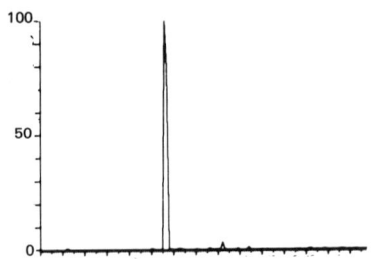

A spectrum was obtained from the probe injection sample number 7 and is shown as the lower spectrum in figure 12. It is seen that a significant ion mass is 327; thus a reconstructed chromatogram from the ion current of mass

Figure 12

SPECTRUM 285-262 FROM LC/MS CHROMATOGRAM

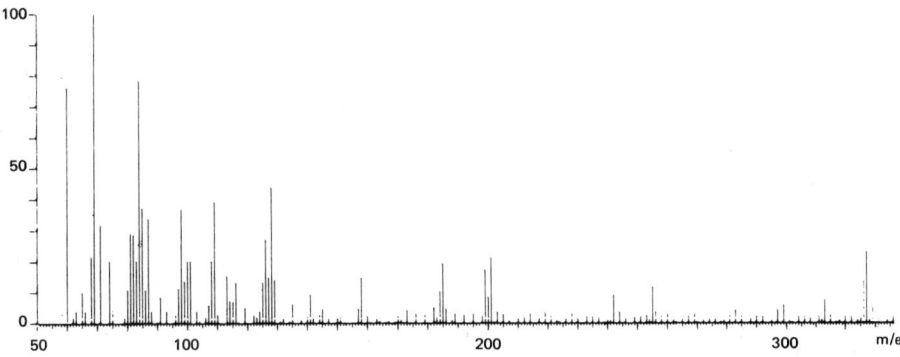

SPECTRUM FROM FRACTION 7

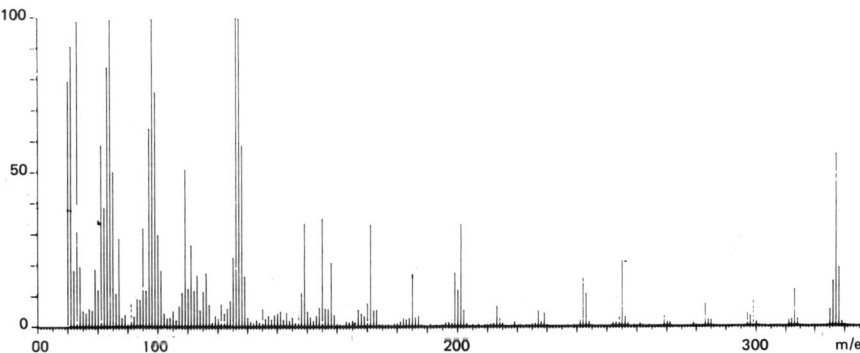

327 was obtained for the probe samples. This is seen as the lower chromatogram on the right-hand side of figure 11. Peak 7 is clearly selected and using the same mass 327 a reconstructed chromatogram was obtained from the original ion-current chromatogram from the LC/MS run. This is shown at the bottom of the right-hand side of figure 11 and again it is seen that the peak is clearly and unambiguously selected from the mixture. A spectrum was then taken from the respective peak in the LC/MS chromatogram and this is shown at the top of figure 12 and can be compared with the spectrum from fraction 7. It is seen that basically the two spectra are identical although the spectrum from fraction 7 obviously contains traces of a contaminating material.

The wire transport system for LC/MS has several distinct advantages. Firstly, as a transport system is employed, its performance is completely independent of the solvent used in the chromatographic system provided it is reasonably volatile. Thus gradient elution development using solvents of any polarity can be employed without affecting the quality of the mass spectra. Secondly, because the transport system is in effect a continuous solid probe injection device, spectra of substances of very low volatility, such as those normally met within liquid chromatographic separations, can be readily obtained. Thirdly, the spectra obtained can be electron impact spectra, which, in general, are far more informative for structure elucidation than spectra obtained by chemical ionization. It should be pointed out, however, that the sensitivity of the wire transport LC/MS system is only moderate but with some development, sensitivities of 10^{-7}-10^{-8} g/ml should be obtainable.

The Finnigan LC/MS Transport System

An LC/MS system based on the device developed by Scott et al. (9,10) is now commercially available and manufactured by Finnigan Inc. The Finnigan instrument, however, utilizes a continuous band or ribbon as the transport medium to increase the quantity of eluent sampled from the column and thus increase the overall sensitivity. Chemical ionization is also employed as an alternative to electron impact ionization to improve sensitivity, but does so at the expense of limiting the structural information obtained from the spectra relative to that obtained from electron impact spectra. The design and performance of the instrument was described by McFadden, Schwartz and Evans (13) and a diagram of their system is shown in figure 13. LC effluent is taken up on the stainless-steel ribbon (3.2 mm wide, 0.05 mm thick) and transported to the vac-locks. Partial evaporation which occurs prior to passage through vac-lock No. 1 can be aided by a combination of heat, gas flow or vacuum as desired. Removal of solvent is completed in vac-locks No. 1 and No. 2 so that less than 10^{-7} g/sec of solvent enters the

Figure 13

The Finnigan LC/MS Transport System

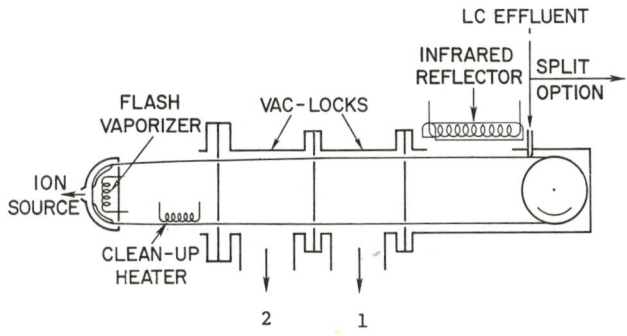

mass spectrometer. Vac-lock No. 1 is pumped with a 500 l/min forepump equipped with an oil mist eliminator and vac-lock No. 2 is pumped with a 300 l/min forepump. Depending on the tolerances set for the vac-lock interface pieces, these chambers are maintained in the pressure ranges 1-20 Torr (vac-lock No. 1) and 0.1-0.5 Torr (vac-lock No. 2) so that the mass spectrometer analyzer section can be pumped to a satisfactory level around 10^{-6} Torr.

Flash vaporization of the sample occurs by radiant heating in a small chamber butted directly to the solid probe entrance of the mass spectrometer ion source. Heat input is provided by a Nichrome heater contained in a quartz tube. The belt travel distance through the chamber is 6.5 cm, so that for most belt speeds (2-4 cm/sec), point residence time is 2-3 sec. By comparison with the temperature of vaporization for solid probe samples, it appears that the belt temperature rises to within 20-30° of the chamber temperature.

The slot in the interfaces for passage of the belt is formed by two "L"-shaped sapphire pieces which are attached to the stainless-steel flange or vacuum closure bar by epoxy cement. The belts used are either 0.05 or 0.075 mm thick and the slot tolerance is set to be 0.075 mm greater than the belt thickness (i.e., either 0.125 or 0.15 mm). The belt width is 0.317 cm and the slot width is 0.325 cm. A ribbon 0.32 cm wide traveling at a speed of 2.5 cm/sec will carry away a liquid film 0.2 mm thick from a solvent flow of 1 ml/min and if the solvent film can be evaporated without loss of solute, then the ribbon will transport virtually 100% of the solute into the mass spectrometer. Sample utilization will then depend only on the efficiency of the flash vaporization step. In practice, some sample is lost by spray processes and the flash vaporization cannot be fully efficient for all compounds. Nevertheless, yields in the range of 25-40% have been attained with an LC/MS ribbon interface system. It follows that the quantity of column eluent taken from the ribbon will be twenty times greater than the wire and provide significantly improved sensitivity.

Figure 14

The Finnigan LC/MS Interface and Ion Source

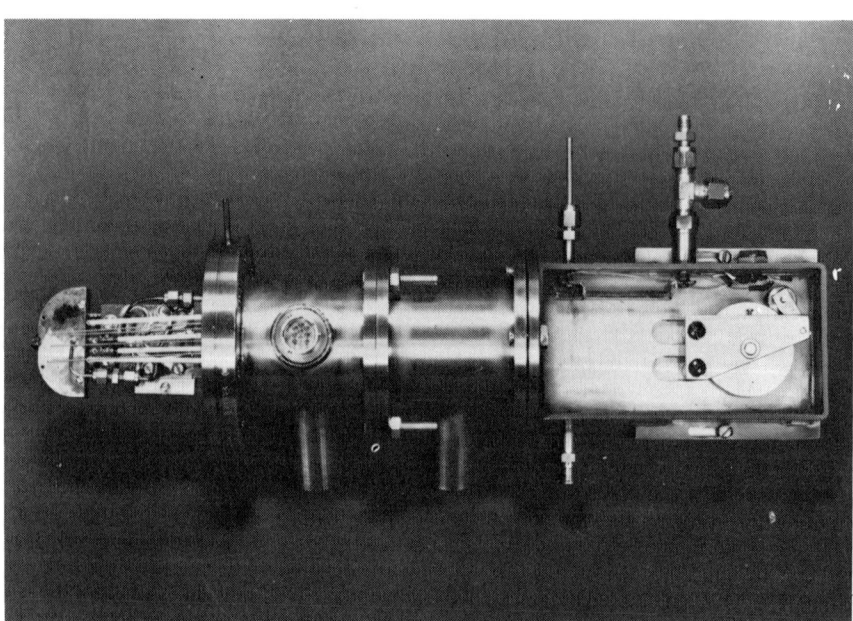

Efficiency of solvent removal through the two vacuum locks is very high, particularly for lower boiling solvents such as hexane. With approximately 10^{-2} g/sec (1 ml/min) of hexane flowing on to the belt, the hexane mass spectral background indicates that around 10^{-7} g/sec is entering the ion source. The enrichment of sample/solvent is therefore in the range of 10^5. Lower boiling solvents such as pentane or methylene dichloride give higher enrichment. Higher boiling solvents such as toluene, dioxane, or isooctane should be avoided since they require heat input for efficient solvent evaporation.

With the current apparatus, optimum performance could be attained with a solvent flow rate of up to 0.85 ml/min. At a higher solvent flow rate, sample is lost due to spray evaporation at the first vacuum lock.

A photograph of the belt interface system and ion source is shown in figure 14. An example of the use of the belt interface system when used to

Figure 15

Chromatogram and Mass Spectra from the Finnigan LC/MS Instrument

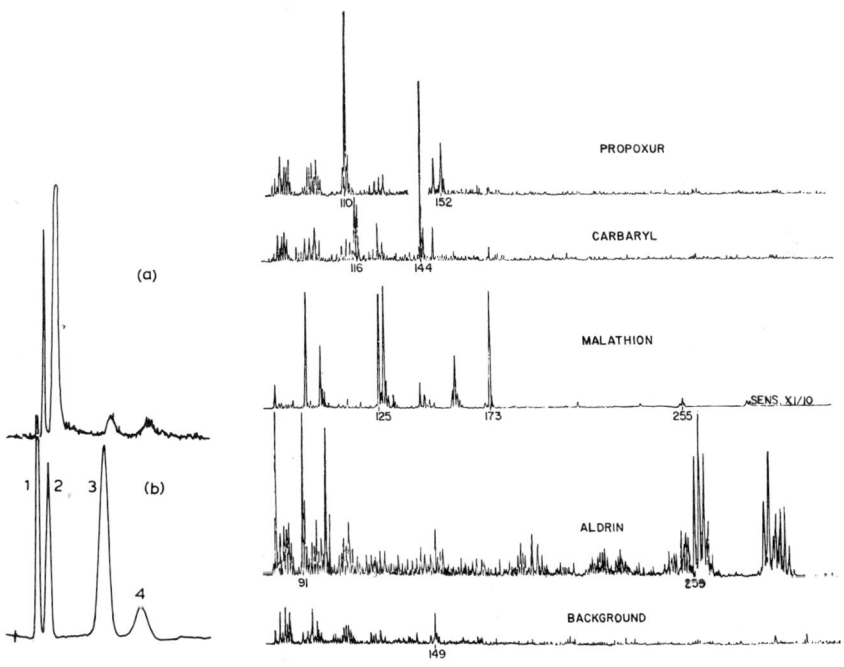

monitor the separation of a pesticide mixture is shown in figure 15. In chromatogram (a) the separation depicts the total ion current monitored by the mass spectrometer during development and chromatogram (b) was obtained by simultaneous monitoring with a UV detector at 254 nm.

Reasonable chromatographic fidelity is maintained during MS detection and very little cross contamination occurs of one compound with adjacent neighbors. The mass spectra obtained at the top of each peak are also shown in figure 15. (Oscillographic traces rather than computer-generated bar graphs are presented here in order to show more realistically the actual quality of spectra produced in the LC/MS mode.) The sensitivity of the system for carbanyl obtained by monitoring the peak mass of 144 was claimed to be 1 ng but insufficient data was given to convert this to the more significant units of g/ml in the column eluent.

The LC/MS system has been demonstrated to be effective, sensitive and useful in practice but it is still a relatively new technique and few laboratories have, at this time, operating LC/MS equipment. However the real value of this technique can only be accurately assessed after a number of laboratories have extensively evaluated the system and demonstrated its efficiency in solving real structural problems.

REFERENCES

1. M. S. Denton, T. P. DeAngelis, A. N. Yacynyck, W. R. Heineman and T. W. Gilbert, Anal. Chem., 48 (1976) 20.
2. Y. Talmi, Anal. Chem., 47 (1975) 658A.
3. Y. Talmi, Anal. Chem., 47 (1975) 697A.
4. D. A. Yates and T. Kuwana, Anal. Chem., 48 (1976) 510.
5. M. J. Milano, S. Lam and E. Grushka, J. Chromatog., 125 (1976) 315.
6. M. A. Baldwin and F. W. McLafferty, Organic Mass Spectrometry, 7 (1973) 1111.
7. P. Arjino, M. A. Baldwin and F. W. McLafferty, Biomed. Mass Spectrom., 1 (1974) 80.
8. P. Arjino, B. G. Dawkins and F. W. McLafferty, J. Chromatog. Sci., 12 (1974) 574.
9. R.P.W. Scott, C. G. Scott, M. Munroe and J. Hess, Jr., The Poisoned Patient: The Role of the Laboratory, Elsevier, New York (1974) 155.
10. R.P.W. Scott, C. G. Scott, M. Munroe and J. Hess, Jr., J. Chromatog., 99 (1974) 395.

11. E. C. Horning, D. I. Carroll, I. Dzidic, K. D. Haegele, M. G. Horning and R. N. Stillwell, J. Chromatog., 99 (1974) 13.
12. R.P.W. Scott and P. Kucera, J. Chromatog. Sci., 11 (1973) 83.
13. W. M. McFadden, H. L. Schwartz and S. Evans, J. Chromatog., 122 (1976) 389.

INDEX

active filters	45
adsorption in detector cell effects of	34
ambient temperature, effect on precision	189
amplifier adjustment	201
analog integration	50
ancillary equipment	41
Beer's Law for light absorption	109
bubbles, their removal and prevention	202
bulk property detectors	55
limitations of sensitivity	56
performance characteristics	57
cell, effect of geometry on dispersion	25
chain detector	169
choice of detector	182
classification of detectors	3
column temperature, effect on precision	190
computer data processing	193
conductivity detector, electrical see electrical conductivity detector	
conductivity detector, practical hints on operation	206
criteria of performance	5
density balance detector	98
density detector	91
detector characteristics that affect column performance	21
detector criteria, summary	37
detector selection	182
dielectric constant detector	70
theory	70
bridge circuits for	72
design of	71
differential mode of detector operation	209
application	211
calibration curves	213
digital integrator	50
diode array detectors	225
examples of use	226
spectra from	227
direct sampling method for LC/MS	230
interface for	231
applications	232
disc detector	178
dispersion, effects of detector	22
drift, removal of	205
Du Pont fluorescence/adsorbance detector	126
optical system	127
specifications	128
dynamic range	13
electrical conductivity detector	80
theory	80
bridge for use with	82
commercial example of	83
specifications of	86
low volume cell	87,88
electromechanical integration	49
electron capture detector	161
specifications	164
fluorometric detector	121
fluoromonitor, optical system	123
specifications	124
function of detectors	1
heat of adsorption detector	139
cell design	140
theory of	143
history of detectors	1
integral mode of detection	215
example of	216
integration, mechanical analog	49
analog	50
digital	50
interferometer detector	95
Laboratory Data Control conductomonitor	83
fluoro monitor	122
spectromonitor I	114
spectromonitor III	113
LC/UV spectrometric systems	224
diode array detectors	225
examples of use	226
spectra from	227
Variscan system	227
optical system	228
LC/MS spectrometer system	229
direct sampling method	230
interface for	231
application	232
the wire transport system	233
interface for	235
operating condition	237
application of	238
sample spectra	239
Finnigan system	240
interface for	241
application of	243
linearity of detector	9
long term noise, reduction in practice	205

measurements, manual and by computer	192	selection of detector	182
moving wire detector, practical hints on operation	205	sensitivity	
		definition	18
nature of detector, output	6	determination of	19
noise		short term noise, removal of	204
effect on precision	194	solute property detector	105
measurement of	17	solvent composition, effect on precision	188
types of	15	solvents	
noise filters		cleaning	105
active, cut-off data	48	properties of	107
example of use	47	special detector techniques	209
passive	45	specific detectors (definition)	105
non linearity, effect on quantitative analysis	10	spectromonitor I	115
		optical system	116
passive filters	46	spectromonitor III	118
performance criteria of detectors	5	spectroscopic detectors	223
		spray impact detector	149
polarographic detectors	131	design	150
design	133	response characteristics	151
electrode potential effect of	155	spurious peaks, their cause and removal	203
precision of chromatographic data		temperature compensated UV detector	110
effect of solvent composition	188	thermal conductivity detector	93
		time constant, amplifier effect on dispersion	27
effect of ambient temperature	189	transport detector, definition	105
effect of column temperature	190	design of	167
		transport LC/MS system	233
effect of sample load	191	interface for	235, 241
qualitative analysis	195	operating conditions	237
quantitative analysis	197	application of	238
radioactivity detector	157	sample spectra	239
cell design	159	turbulence, effect on dispersion	26
reactor		UV detector	109
characteristics	43	temperature compensation	110
design of	44	effect of temperature control on noise	111
predetector	41		
recorder, adjustment	201	UV/LC system, see LC/UV system	
recorder time constant	33	vacancy chromatography	217
refractive index detector		apparatus for	219
angle of deviation method	59	examples of	220
critical angle method	60	variable wavelength UV detector	114
Fresnel method	60	Variscan LC/UV system	227
Christiansen method	62	optical system	228
Waters refractometer	63	vapor pressure detector	99
application	67	Waters differential refractometer	63
practical operating hints	207	specifications	66
response	14	wire transport detector	
different types of	7	by pyrolysis	168
response index		by methane conversion	171
definition of	9	linearity of	173
determination of	11	specifications	174
sample load, effect on precision	191		

OHIO UNIVERSITY LIBRARY

Please return this book as soon as you have finished with it. In order to avoid a fine it must be returned by the latest date stamped below.

APR 4 1980
SEP 8 1993
JUN 2 1980
QUARTER LOAN
DEC 24 1980
AUG 23 1993
JAN 27 1981
~~FEB 17 1981~~
FEB 17 1981
APR 30 1981
QTR. LOAN
JAN 17 1983
MAR 15 1983
QTR. LOAN
JUL 23 1983
AUG 26 1983
QTR. LOAN
FEB 3 1984
APR 2 1984
QTR. LOAN
JUL 5 1985
CF SEP 11 1985